PLANT SCIENCE RESEARCH AND PRACTICES

ORCHIDACEAE

CHARACTERISTICS, DISTRIBUTION AND TAXONOMY

PLANT SCIENCE RESEARCH AND PRACTICES

Additional books and e-books in this series can be found on Nova's website under the Series tab.

PLANT SCIENCE RESEARCH AND PRACTICES

ORCHIDACEAE

CHARACTERISTICS, DISTRIBUTION AND TAXONOMY

VLADAN DJORDJEVIĆ, PHD
EDITOR

nova science publishers
New York

https://doi.org/10.52305/LWER2262

Library of Congress Cataloging-in-Publication Data

ISBN: 978-1-68507-189-9

Published by Nova Science Publishers, Inc. † New York

Contents

PREFACE

The orchid family (Orchidaceae) is one of the best-studied of plant families, known for its great species richness, cosmopolitan distribution, and rarity. The distinctiveness of orchids is mainly reflected in their complex biology, i.e., the relationships they form with mycorrhizal fungi and pollinators. In addition, orchids fascinate scientists and nature lovers because of their origin and speciation, their often exotic flowers, the large number of species of orchids, and the peculiarities of their life history. Many orchid species are rare, threatened, or endangered and are under legal protection in many countries. Today there is increasing interest among scientists in questions of taxonomy, ecology, distribution, and conservation of both epiphytic and terrestrial orchids. This book is a journey through our knowledge about orchids, starting with orchid fossils, age issues, and study of extinction rates of orchids, moving on to an overview of the national orchid flora of less explored regions, chorological contributions, and ecological studies, and finally covering developments in orchid breeding through genetic engineering. The main objective of the book is to familiarize the reader with the basic features of the family Orchidaceae from various aspects. Eight chapters by expert authors cover the characteristics, distribution, ecology, and taxonomy of orchids. The book is intended for scientists working on orchids and researchers interested in botany and conservation, as well as biologists and ecologists in general, graduate students, scholars, and nature lovers.

Chapter 1 represents a review of the fossil remains of orchids in Tertiary (Cenozoic) amber from the Baltic region, Mexico, and the Dominican Republic. The author notes that amber contains the remains of orchid flowers, orchid seeds, and insects with attached orchid pollinia, and emphasises that amber provides information about a particular geological time period when certain features were established, but also about the geographical location where these features arose.

Chapter 2 deals with extinction rates of dominant orchid species during the last 150 years in the Czech Republic. The authors analyze existing trends in changes in the number of sites of abundant orchid species and put the available data into context with shifts in agricultural practices. Their conclusion is that species with many historically recorded sites suffer a very low site extinction rate, which may be because these species are widely represented and do not have great demands in regard to site quality. Species with few historically recorded sites suffer higher site extinction rates, which may be because of their special site quality requirements. The site extinction rate in species with many historically recorded sites was largest during the communist era (1948-1989). From the conservation point of view, this means that extinction-prone species should be paid special attention through constant maintenance of good conditions of their sites, while site quality is not so important for more abundant species, as extinction is less likely.

Chapter 3 presents an overview of the orchid flora of Serbia, a country of Southeast Europe, taking into account its taxonomic structure, life forms, and pollination systems. In addition, a detailed insight into the chorological groups of each orchid taxon is given. A total of 72 orchid species and subspecies belonging to 22 genera have been recorded in Serbia, whereas the presence of three taxa is doubtful. A specific feature of the orchid flora of Serbia is the presence of two Balkan endemics (*Himantoglossum calcaratum* subsp. *calcaratum* and *Dactylorhiza cordigera* subsp. *bosniaca*) and three subendemics of the Carpathians and the Balkans (*Gymnadenia frivaldii*, *Dactylorhiza maculata* subsp. *transsilvanica*, and *Dactylorhiza cordigera* subsp. *cordigera*). This synthesis underlines the fact that the study area provides suitable conditions for the survival of some orchid species that have southern or northern limits of their distribution in this part of Southeast Europe.

Chapter 4 deals with the distribution and conservation status of orchids of the Republic of Mordovia (Russia). According to the summarized and standardized data, 24 orchid species are recorded in Mordovia. The hotspots of orchid diversity in the study area are pointed out, and the problems in orchid research and possible ways to solve them are highlighted.

Chapter 5 aims to explore the relationship between the community of orchid species and their associated phorophytes in Cabo San Antonio, Guanahacabibes National Park, Cuba. The results show that the relationship between the epiphyte community and phorophyte species is generally asymmetric and such that there are generalist epiphyte species with low habitat specificity or requirements, and phorophytes that are also generalists and can host several orchid species and other epiphytes. Some orchid species were also found only on specific phorophyte species, even in areas where other phorophyte species were present nearby. The authors hypothesize that orchids may have specificity for certain types of mycorrhizal fungi required for seed germination that are found only on the bark of these phorophyte species.

Chapter 6 analyzes some structural features of a population of *Encyclia bocourtii*, an epiphytic orchid endemic to the Guanahacabibes Peninsula (Cabo San Antonio) in western Cuba. This species showed an aggregate distribution pattern with high co-specific competition along the five transects studied. It was recorded selectively on only three species of phorophytes (*Chrysobalanus icaco*, *Metopium brownei*, and *Thrinax radiate*). *Encyclia bocourtii* was found to have a metapopulation structure, with the highest density occurring in coastal vegetation. The hypothesis of site quality is discussed in relation to the possible presence of mycorrhizae.

Chapter 7 aims to review and analyze the characteristics of *Epipactis helleborine* (L.) Crantz, one of the most widespread orchids of the Temperate Zone, that allow it to grow in different kinds of habitats. The authors point out that the potential of *E. helleborine* to occupy new habitats and spread is due to its enormous phenotypic plasticity, as well as to a wide range of mycorrhizal symbionts and pollinating insects. Thanks to its remarkably high adaptability to various environmental conditions, this taxon can be classified as an "undemanding species". In fact, such a configuration of characteristics makes it a "super species" that will surprise us with the conquest of new habitats and territories in the future.

Chapter 8 points out that genetic engineering is likely to play a very important role in the development of new orchid varieties in the future. At the same time, this chapter stresses that the discovery, isolation, and identification of orchid flower genes, understanding the principles of genomics, and study of functional genomics will lead to greater applicability of genetic transformation. Gene transfer technology in orchids has the potential to change a particularly desired set of floral traits via genetic engineering of a single gene or stack of genes, thereby overcoming the various limitations of traditional breeding methods. New molecular techniques enable the breeder to select and regenerate desired orchid plants in a much shorter time. In the future, these genetic transformation technologies can be used to target positive traits like disease and pest resistance, cold hardiness, adjustment of flowering time according to market demand, etc., and thus be of great benefit to producers. The genome editing technique can be successfully used for precise modification of the genome through genetic transformation to ultimately generate novel genotypes of orchids.

Overall, the book "Orchidaceae: characteristics, distribution, and taxonomy" provides a rich background for future research on orchids from a variety of perspectives. The findings presented in the book indicate that future research should not neglect classical basic research. Given the complexity of the biology and ecology of taxa within the family Orchidaceae, it is clear that the multidisciplinary approach to research needs to be emphasized. The recommendations that emerge from each chapter of this book point to a new era in the study of this fascinating plant family.

In: Orchidaceae
Editor: Vladan Djordjević
ISBN: 978-1-68507-189-9

Chapter 1

The Fossil Record of Orchids in Amber

George Poinar Jr.*
Department of Integrative Biology, Oregon State University, Corvallis, OR, USA

Abstract

Surprisingly, the fossil record of orchids is not extensive considering that the family is one of the most specious of the Angiosperms. This is partly because there are so very few possibilities for orchid remains to be fossilized, especially in the subtropics and tropics where the majority of orchids are found. While compression fossils of orchid plants occur, certainly the most eloquently preserved orchid remains are found in amber. This is due to the ability of plant resins to infiltrate, fix, and dehydrate embedded plant tissues. Also since amber contains a wide range of insects, it provides a window to a wide range of pollinia that are attached to various pollinating insects, including those of the orders Hymenoptera, Diptera and Coleoptera. In fact, the oldest known orchid fossil is a pollinarium attached to a 40-45 million year old Baltic amber fungus gnat. The preservative qualities of

* Corresponding Author's E-mail: poinarg@science.oregonstate.edu.

amber also make it possible to study orchid seeds, even along with their symbiotic fungus, providing evidence that this partnership had already been established some 15-40 MYA. Indirect fossil evidence of orchids also occurs in amber with the preservation of insects known to pollinate extant orchids, such as tropical orchid bees. A review of the above fossil associations in amber is presented below, along with accompanying figures.

Keywords: Tertiary, amber, pollinators, pollinia, seeds

Introduction

The present work reviews the fossil remains of orchids in Tertiary (Cenozoic) amber from the Baltic Region, Mexico and the Dominican Republic. Orchid fossils and subfossils in sedimentary deposits have been adequately reviewed previously (Schmid and Schmid 1977; Conran et al., 2009; Gołaszewska et al., 2019). Plant remains in amber are often eloquently preserved since plant resins have the ability to infiltrate, fix, and dehydrate different stages of plant remains, including those of orchids. Soon after plant material has entered a resin deposit, oxygen is restricted and associated decay-causing microorganisms are destroyed. Terpenoids and acids in the resin fix the plant tissues while sugars and terpines replace water by inert dehydration (Poinar and Hess 1985). Amber contains the remains of orchid flowers, orchid seeds and insects with attached orchid pollinia. By providing these fossils, amber gives us not only a particular geological time period when certain feature were established, but the geographical location where these features occurred.

Amber Deposits

Fossils discussed in the present work originated from three separate amber deposits. The first was Dominican amber collected from mines in the Cordillera Septentrional of the Dominican Republic. While dating of Dominican amber is generally assumed to be 15-20 MYA based on foraminifera (Iturralde-Vinent and MacPhee 1996), a date of 30-45 MYA based on coccoliths was proposed by Cêpek (Schlee 1990). Dominican amber is secondarily deposited in sedimentary rocks, which makes a definite age determination difficult (Poinar and Mastalerz 2000). A range of ages for Dominican amber is possible since the amber is associated with turbiditic sandstones of the Upper Eocene to Lower Miocene Mamey Group (Draper et al., 1994). Such a range of ages was demonstrated by nuclear magnetic resonance spectroscopy (NMR) analyses (Lambert et al., 1985). Using this method, the ages of the various mines ranged from 20 to 40 million years before present (lower Miocene to upper Eocene).

Dominican amber was produced by the leguminous tree, *Hymenaea protera* Poinar (1991), and a re-construction of the Dominican amber forest based on amber fossils indicated that the environment was similar to that of a present day tropical moist Neotropical forest (Poinar and Poinar 1999).

Mexican amber originated from lignitic beds among sequences of marine calcareous sandstones and silt in the northern mountain ranges of Simojovel in Chiapas, Mexico. The amber is associated with Early Miocene Balumtun Sandstone and Late Oligocene La Quinta Formation with radiometric ages from 22.5 to 26 million years (Berggren and Van Couvering 1974). Since the amber is secondarily deposited in these marine formations, it may be somewhat older than the above dates. Mexican amber was produced by *Hymenaea mexicana* (Fabaceae) (Poinar and Brown 2002).

Baltic amber was formed in tropical and subtropical forests that covered a large part of northern Europe for some 10 MYR during the early Tertiary. It was weathered from the original sediment and re-deposited in marine 'blue earth' layers in the Samland Peninsula. Original age estimates of Baltic amber were 35 to 55 MYR (Larsson 1978). The latest estimate, based on geo- chemical tests, is from 45 to 55 MYA (Wolfe et al., 2016). A wide diversity of angiosperms (five families of monocots and 38 families of dicots) has been described from Baltic amber, but none were assigned to Orchidaceae (Czeczott 1961).

Observations and photographs in the present work were made with a Nikon SMZ-10 R stereoscopic microscope and Nikon Optiphot compound microscope with magnifications up to 800 X. Helicon Focus Pro X64 was used to stack photos for better depth of field.

Direct Evidence of Orchid Remains in Amber

Orchid Pollinaria in Amber

Most orchids dispense their pollen in little sacs called pollinia that are attached by supports (caudicles) to adhesive pads (viscidia) attached to various parts of pollinators. The entire unit is known as a pollinarium. Of all the angiosperms, only orchids and milkweeds independently evolved this method of pollination (Endress 1994). Just when orchids began dispensing their pollen in pollinia is unknown, but the great diversity of pollinia shapes and the various ways they are attached to pollinating insects is quite outstanding. All insects carrying orchid pollinaria are considered orchid pollinators (Ackerman and Mesler 1979).

Orchid pollinaria can be used for taxonomic identifications of orchid lineages that produced the pollinaria (Singer et al., 2008). Since pollen dispensed in pollinia is not available as food to most insects, orchids supply nectar or have evolved ingenious patterns of mimicry and deceit to attract potential pollinators (Dressler 1993). Pollinia have been found in amber attached to bees (Hymenoptera), a fly (Diptera) and beetles (Coleoptera).

Pollinaria Associated with Bees

The first record of a pollinarium attached to a bee in amber is *Meliorchis caribea* Ramírez, Gravendeel, Singer, Marshall and Pierce. The pollinarium is fastened to the back of a worker stingless bee (Hymenoptera: Apidae) in Dominican amber. It is exceptionally well preserved and was placed in the subtribe Goodyerianae of the Tribe Cranichidae of the subfamily Orchidoideae (Ramírez et al., 2007).

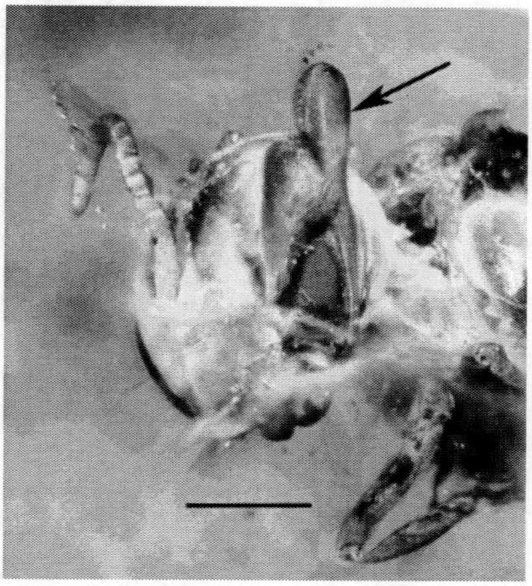

Figure 1. Pollinarium of *Rudiculites dominicana* attached to left eye of stingless bee in Dominican amber. Scale bar = 0.5mm.

Two additional pollinaria were described from the stingless bee, *Proplebeia dominicana* (Wille & Chandler) in Dominican amber. One, described as *Rudiculites dominicana* Poinar (2016A), was attached to the left compound eye of the bee by an elongate viscidium (Figure 1). The 1.2 mm long pollinarium was spoon-shaped with narrow caudicles and a wider area with two pollinia. Pollen clusters could be seen inside the pollinium. Based on similarly structured elongate pollinia on extant members of *Spiranthes* Richare and *Sarcoglottis* C. Presl that have bees as the main pollinators, the fossil was placed in the subtribe Spiranthinae; tribe Cranichideae of the subfamily Orchidoideae (Poinar 2016A). The subtribe Spiranthinae, to which *Rudiculites dominicana* gen. n., sp. n. is tentatively assigned, includes terrestrial and epiphytic orchids with short, leafy stems and rhizomes, fleshy clustered roots and soft herbaceous leaves. The flowers are small to medium with a deep nectary united to the ovary. Most of the 400 some species in 30-40 genera are Neotropical and pollination is normally carried out by bees (Dressler 1993).

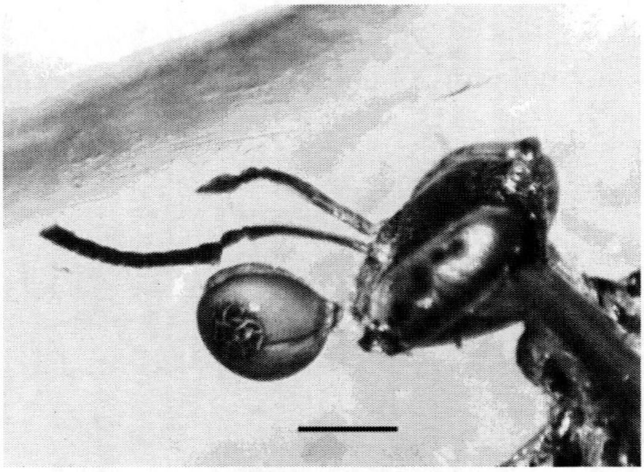

Figure 2. Pollinarium of *Globosites apicula* attached to the mouthparts of a stingless bee in Dominican amber. Scale bar = 0.5 mm.

Another Dominican amber pollinarium was attached to the mouthparts of a second specimen of *Proplebeia dominicana* and described as *Globosites apicula* Poinar (Figure 2). The pollinarium was globose, with two appressed pollinia and a small translucent viscidium attached to the mouthparts of the bee. It was tentatively assigned to subfamily Epidendroideae, tribe Cymbidieae, subtribe Oncidiinae (Poinar 2016A). The 1200 species of the tropical American subtribe Oncidiinae are mostly epiphytes that are pollinated by bees and wasps (Dressler 1993; Poinar 2016A).

Pollinarium Associated with a Fly

A Baltic amber female fungus gnat was discovered to have a detached orchid pollinarium attached to the base of its leg (Figure 3). The pollinarium consisted of two pollinia connected to two viscidia by hyaline caudicles. Some pollen grains remained in an opened pollinium.

This fossil, which presently is the oldest fossil record of an orchid, was described from Baltic amber as *Succinanthera baltica* Poinar & Rasmussen and assigned to the subfamily Epidendroideae (Poinar and Rasmussen 2017).

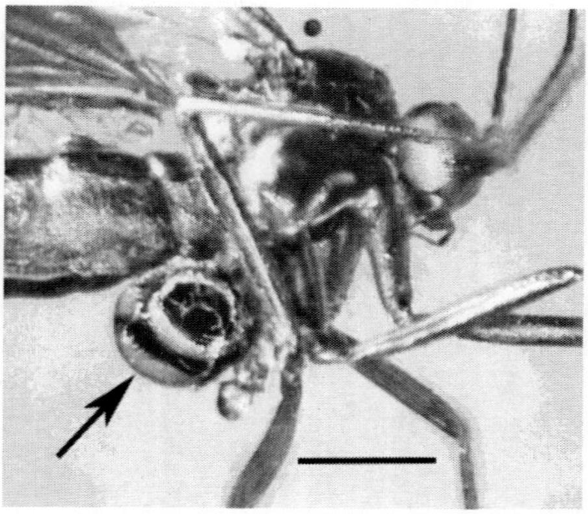

Figure 3. Pollinarium of *Succinanthera baltica* attached to the leg base of a fungus gnat in Baltic amber. Scale bar = 500 μm.

The fly carrier was identified as a species of *Bradysia* Winnertz 1867, an extant genus of the family Sciaridae (Diptera). The pollinarium is attached to the coxa and trochanter of the hind leg. The remainder of the leg is missing and was probably detached when the fly acquired the pollinarium. There are records of extant *Bradysia* spp. that also carry pollinaria attached to the base of their hind legs, similar to the attachment position in the fossil (Duque-Buitrago and Alzate-Quintero Otera 2014). Fungus gnats are also known to be effective pollinators of some extant orchids (Mesler et al., 1980).

Pollinaria Associated with Beetles

Associations between beetles and orchids are quite diverse. Beetles that visit orchid flowers for nectar may acquire pollinaria and serve as pollinators. Some orchids depend on various Coleoptera for pollination and have produced "cantharophilous" flowers especially attractive to beetles (Endress 1994; Kalshoven and van der Laan 1981). Other beetles raise their young on orchid floral and vegetative parts (Poinar 2016B).

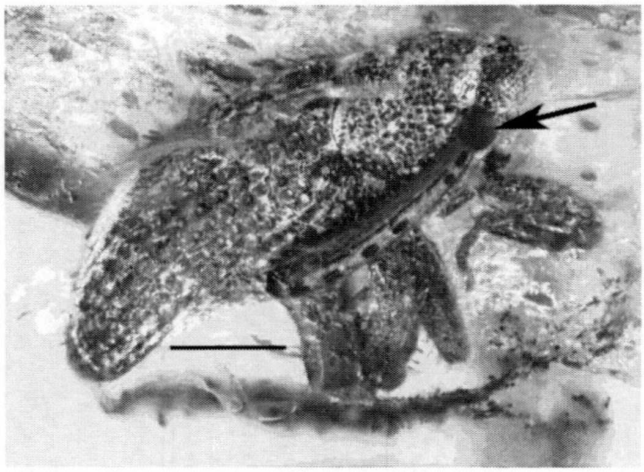

Figure 4. Pollinarium of *Cylindrocites browni* attached to the pronotum of a weevil in Dominican amber. Scale bar = 780 μm.

An orchid pollinaria, described as *Cylindrocites browni* Poinar, was discovered attached to a cryptorhynchid weevil in Dominican amber (Figure 4). The long, cylindrical pollinia resemble those of the extant orchid genera *Dichromanthus* Garay, *Cyclopogon* C. Presl and *Schiedeella* Schltr. of the subtribe Spiranthinae Lindley that occur today in Mexico, the West Indies, and Central America (Poinar 2016B).

Members of the subtribe Spiranthinae are terrestrial or epiphytic orchids with short, leafy stems and rhizomes, fleshy clustered roots, and soft, herbaceous leaves. The flowers are small to medium with a deep nectary united to the ovary. Most of the 400 species in 30–40 genera are Neotropical, and pollination is mostly by various types of bees (Dressler 1993). Since deceit pollination does not occur in the Spiranthinae (Givnish et al., 2015), it is likely that the weevil was attracted by nectar deposits in the flowers of *Cylindrocites browni.* A mucus-like sticky viscidium attaches the pollinia to the anterior portion of the pronotum of the fossil weevil. The weevil is adjacent to one of two floral elements that are considered to belong to the orchid that produced the attached pollinia (see additional discussion below).

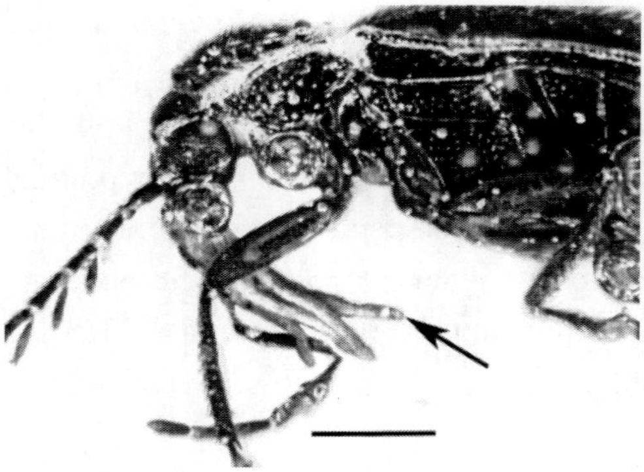

Figure 5. Pollinarium of *Annulites mexicana* attached to the mouthparts of a ptilodactyline beetle in Mexican amber. Scale bar = 300 μm.

While there are no reports of extant cryptorhynchid weevils visiting orchid flowers, pollinia of *Dendrochilum longiflorium* Rchb.f. (subfamily Epidendroideae) have been found attached to the heads of *Rhynchophorus* sp. weevils (Ridley 1896) and pollinia of *Dockrillia linguiformis* (Brieger) (subfamily Epidendroideae) are carried by the eugnomine weevil, *Ancyttalia sydneyensis* (Blackburn) (Zimmerman, 1992). In addition, a number of unidentified weevils carry pollinia of *Pleurothallis* R.Br. (subfamily Epidendroideae) (van der Cingel 2001).

An unusual pollinarium was found attached to a 1.4 mm long ptilodactyline beetle in Mexican amber. The pollinarium consists of four long, annulated pollinia with pollen tetrads arranged in transverse rows. The structure, which has a short visidium and reduced caudicles, is attached to the mouthparts of the beetle. The pollinarium was described as *Annulites mexicana* Poinar (2016B) and assigned to the subtribe Cranichidinae based on the resemblance of the elongate segmented pollinia to those of representatives of the extant genus *Solenocentrum* Schltr. Flowers of this subtribe are small to medium and borne on short stems bearing soft leaves. It is assumed that the various segments of *Annulites*, which bears pollen tetrads between them, can fracture and release the pollen.

Similar pollinaria occur in species of the extant genus *Solenocentrum* (Poinar 2016B).

No extant ptilodactylids are known to carry pollinaria so it is difficult to determine if the fossil species is a chance pollinator searching for nectar in orchid flowers or a specific associate using the orchid for food and development of its young. Since deceit pollination has not been reported for this orchid subtribe (Givnish et al., 2015), the beetle may just have been obtaining nectar from the flowers of *Annulites mexicana.*

Orchid Floral Remains in Dominican Amber

A cryptorhynchine weevil with an attached pollinarium described as *Cylindrocites browni* is adjacent to floral remains (length = about 16 mm, greatest width = about 8 mm) in a piece of Dominican amber (15-45 million years old) (Poinar 2016B). The floral remains (Figure 6) possess features of *Dichromanthus* and *Cyclopogon*, two genera of the subtribe Spiranthinae that possess features of the pollinarium carried by the adjacent weevil. These genera are found today in Mexico, the West Indies and Central America. (Poinar 2016B).

Using micromorphology, the surfaces of the floral remains revealed many, elongate, unicellular trichomes, some of which appeared to contain oil deposits (Figure 7). The trichomes were unicellular and unbranched with the bases attached to the epidermis of the floral remains. The majority of the long, slender trichomes range from 341 μm to 496 μm in length (average = 415 μm; N = 10) and 12 μm - 31 μm in greatest width (average of 22 μm; N = 10). Attached cystoliths ranged from 10 μm to 16 μm in greatest diameter. What appeared to be nectar or oil droplets within the trichomes ranged from 38 μm to 45 μm in greatest diameter. Nectariferous and oleiferous trichomes are common components of orchid flowers, especially those that require insects to carry their pollinaria (Stpiczyńska et al., 2018).

Figure 6. Floral remains, together with a weevil (arrow) bearing a pollinarium of *Cylindrocites browni*, in Dominican amber. Scale bar = 3.6 mm.

Some edges of the floral remains bear germinating pollen grains and pollen tubes (Figure 8), reminiscent to those found on transmitting tracks of orchid flowers. These floral remains help to complete the description of *Cylindrocites browni*, which previously was only known from the pollinarium attached to its beetle pollinator (Poinar 2016B).

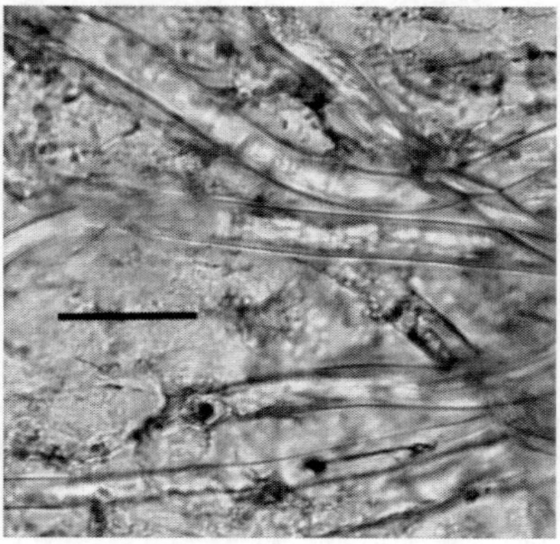

Figure 7. Trichomes containing possible oil deposits on floral remains of *Cylindrocites browni* in Dominican amber. Scale bar = 170 μm.

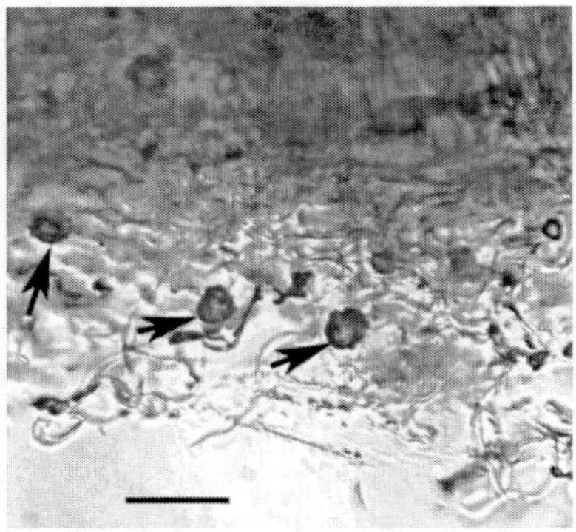

Figure 8. Pollen grains (arrows) and pollen tubes adjacent to a likely transmitting track of *Cylindrocites browni* in Dominican amber. Scale bar = 43 μm.

Putative Orchid Capsule in Dominican Amber

A very small, 2-carpelled, narrow capsule (16.5 mm in length and 172 μm in width), with a thin seed coat having longitudinal thickenings 9-11 μm thick, was discovered in Dominican amber (Figure 9). Removal of a portion of the seed coat revealed numerous microscopic seeds inside (Figure 10). The flat, elongated seeds were 20-23 μm in length, 7-9 μm in width and possessed a central nucleus (Figure 11).

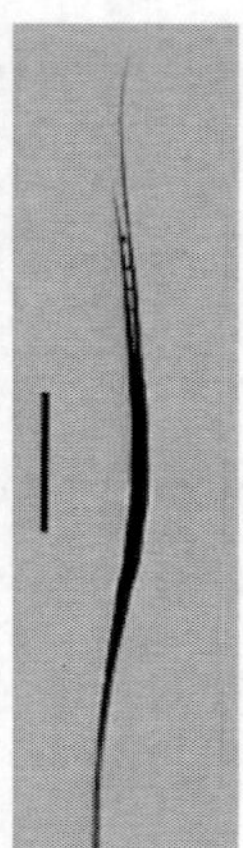

Figure 9. Small narrow capsule in Dominican amber. Scale bar = 3 mm.

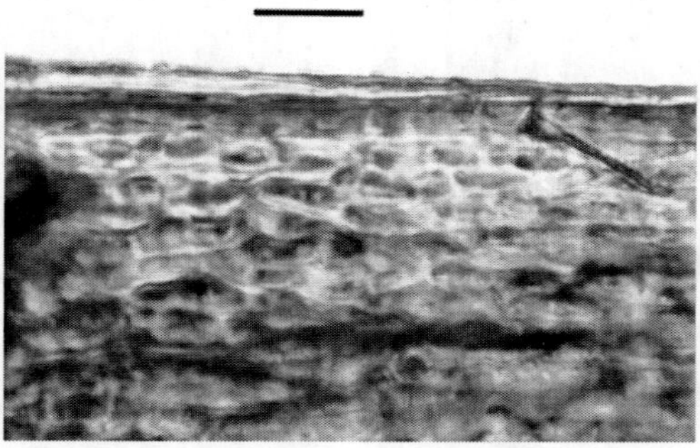

Figure 10. Section of capsule with dust seeds packed inside in Dominican amber. Scale bar = 17 μm.

While most orchid capsules have 3 carpels or valves, those of the genera *Dichaea*, *Vanilla* and *Pleurothallis* can split into two unequal valves, as in this Dominican amber capsule (Dressler 1993).

While the structure resembles a miniature *Vanilla* capsule, the seeds do not have the typical shape of that genus (Figure 11).

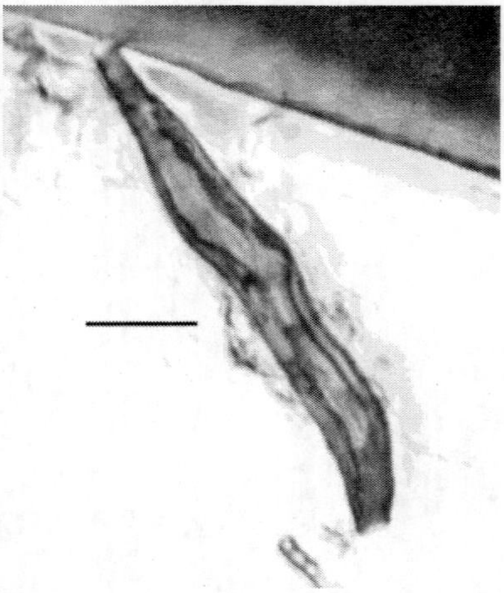

Figure 11. Individual seed from the capsule in Dominican amber. Scale bar = 5 μm.

Orchid Seed in Dominican Amber

Of special interest is a germinating orchid seed, described as *Mycophoris elongatus*, together with its fungal symbiont, *Synaptomitus orchiphilus*, in Dominican amber (Poinar 2017A,B) (Figure 12).

The extended testa cells contained evidence of a symbiotic fungus (Figure 13) and the basal sector showed what appear to be developing rootlets and a suspensor (Figure 14).

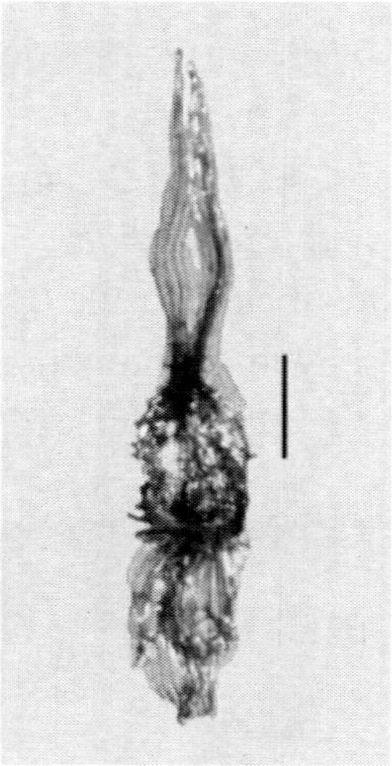

Figure 12. Germinating seed of *Mycophoris elongatus* in Dominican amber. Scale bar = 312 μm.

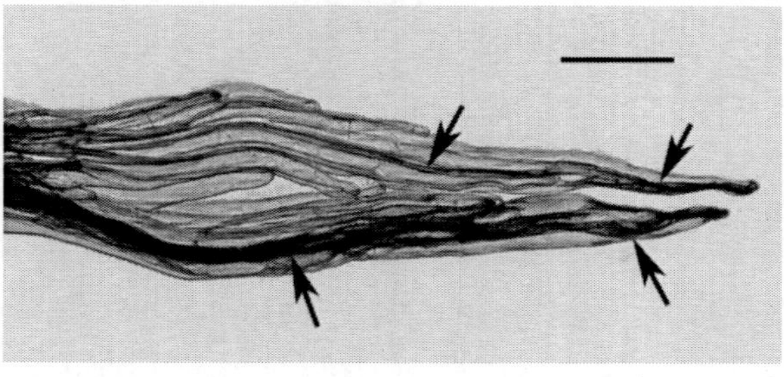

Figure 13. Extended testa cells of *Mycophoris elongatus* in Dominican amber. Arrows show regions with symbiotic fungi. Scale bar = 98 μm.

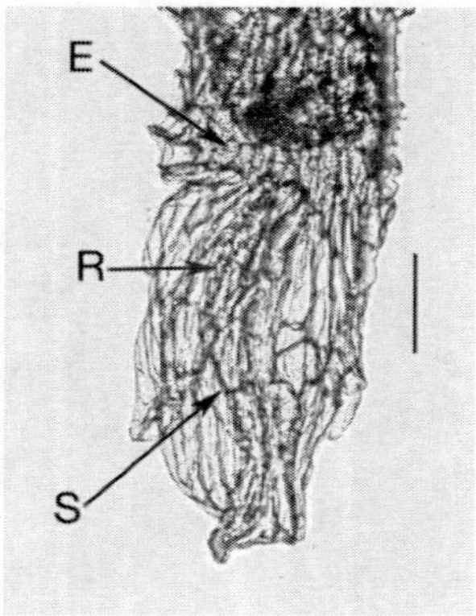

Figure 14. Basal sector of *Mycophoris elongatus* in Dominican amber. E = dividing embryo cells; R = possible rootlet; S = Suspensor. Scale bar = 30 µm.

The conclusion that this is an orchid seed, rather than another type of dust seed, was based on the following features of the fossil.

1. The length of the fossil seed (1.3 mm) is well within the range of most orchid seeds (which range from 0.05 mm to 6.0 mm in length) (Arditti and Ghani 2000).
2. The spherical or globular shape of the embryo (Figure 15) and the shape of the seed coat.

Figure 15. Spherical embryo (arrow) of *Mycophoris elongatus* in Dominican amber. Scale bar = 111 µm.

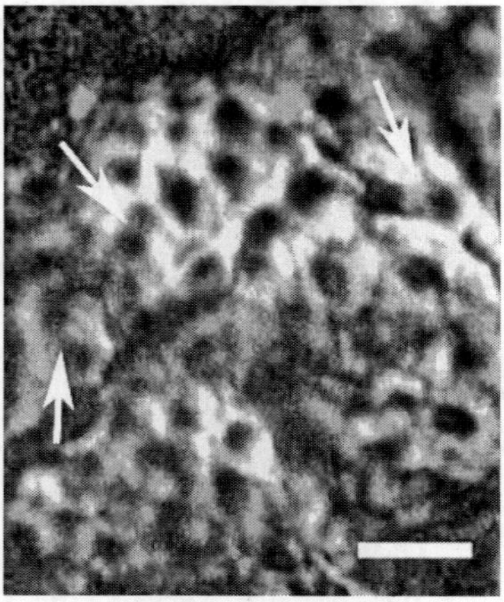

Figure 16. Dividing embryo cells (arrows) of *Mycophoris elongatus* in Dominican amber. Scale bar = 14 μm.

These parameters can vary greatly in extant orchids, with seeds ranging from elliptical to globular, angular or rounded, long or short, transparent, translucent or opaque (Arditti 1967).

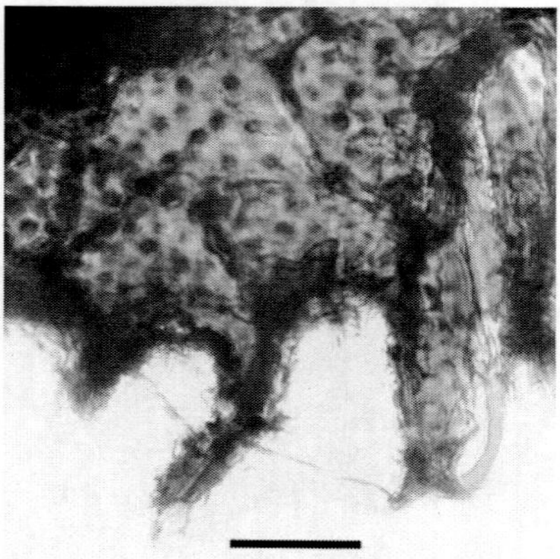

Figure 17. Developing rhizoids of *Mycophoris elongatus* in Dominican amber. Scale bar = 44 μm.

3. Dividing embryo cells with distinct nuclear divisions. These are characteristic of orchid embryos infected with symbiotic fungi (Arditti 1967) (Figure 16).
4. Developing rhizoids showing that *M. elongatus* was in an early stage of protocorm formation (Figure 17).

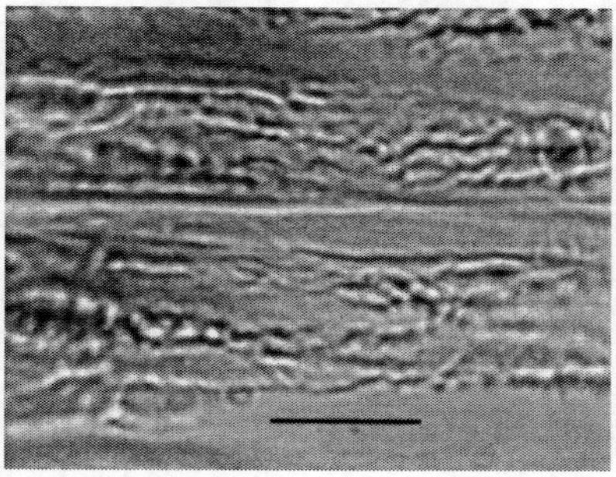

Figure 18. Hyphal strands of *Synaptomitus orchiphilus* in Dominican amber. Scale bar = 23 μm.

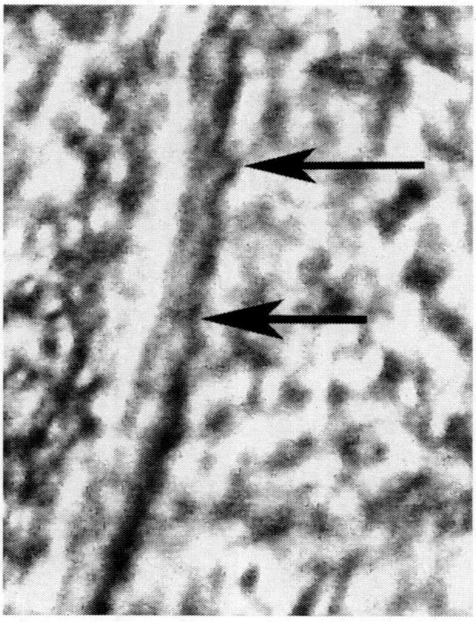

Figure 19. Clamp connection (top arrow) and cell wall (bottom arrow) of *Synaptomitus orchiphilus* in Dominican amber. Scale bar = 11 μm.

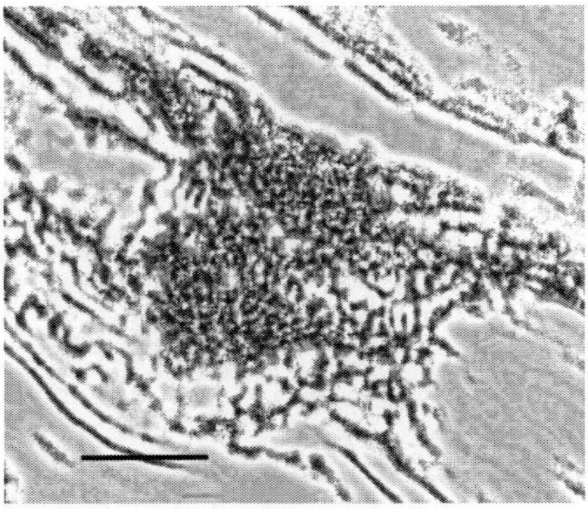

Figure 20. Peloton of *Synaptomitus orchiphilus* in Dominican amber.
Scale bar = 8 μm.

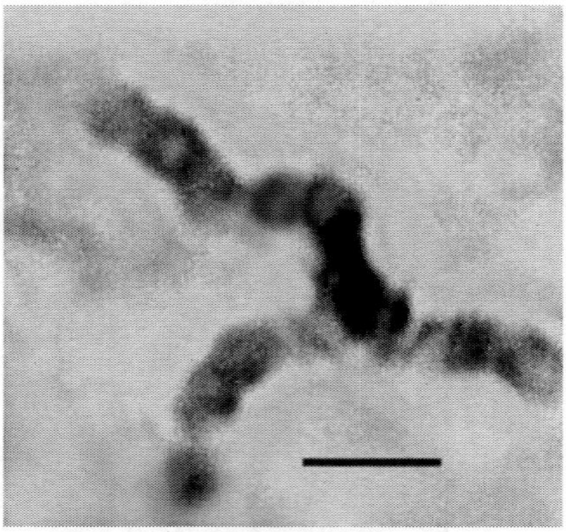

Figure 21. Monilioform cells of *Synaptomitus orchiphilus* in Dominican amber.
Scale bar = 36 μm.

Features of the associated symbiotic fungus, *Synaptomitus orchiphilus*, which are characteristic of orchid mycorrhiza, are: 1. Hyphae with cross walls and clamps, which are found in the great majority of germinating orchid seeds (Arditti 1967) (Figures 18, 19): 2. Evidence of penetration of the fungus through suspensor cells of the seed (Figure 14), which is the usual manner of entry of symbiotic fungi into developing orchid seeds (Arditti 1967): 3. The presence of pelotons (intracellular hyphal coils) (Figure 20) and monilioform cells (Figure 21) inside orchid embryo cells: 4.

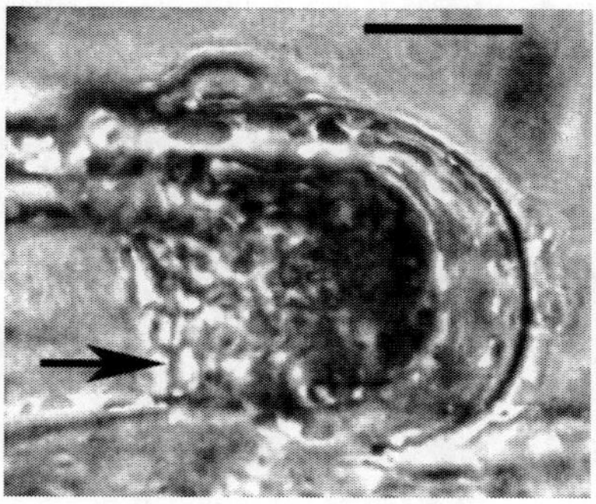

Figure 22. Area of basidia (arrow) with spores of *Synaptomitus orchiphilus* in Dominican amber. Scale bar = 34 μm.

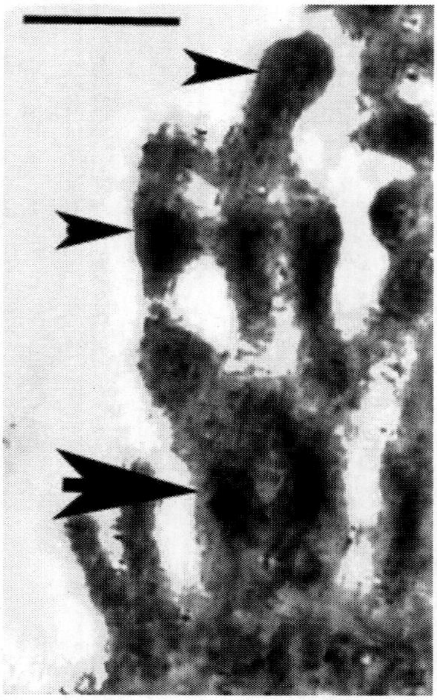

Figure 23. Detail of basidium (large lower arrow) and basidiospores (upper smaller arrowheads) of *Synaptomitus orchiphilus* in Dominican amber. Scale bar = 15 μm.

The formation of basidia and basidiospores within the seed (Figures 22, 23). While not all of the basidia of *Synaptomitus* bear 4 basidiospores, extant Agaracales can develop basidia with one, two or three sterigmata. There are an extensive number of basidiomycetes exhibiting heterospory (Bauch 1926; Halbwachs and Bassler 2015).

The characters presented above, especially the teleomorphic stages of *Synaptomitus orchiphilus* with the formation of basidia and basidiospores inside embryo cells of *Mycophoris elongatus*, support the opinion that *Mycophoris elongatus* represents an orchid seed and *Synaptomitus orchiphilus* is its mycorrhizal fungus.

The seed identification and presence of a symbiotic fungus were questioned by Selosse et al., (2017). Remarks made in the latter paper clearly show that biologists with no experience with fossils, especially amber inclusions, do not understand that interpreting diagnostic characters in fossils can be quite different from examining comparative stained, extant material. In microscopic fossils, size and clarity of features are often modified due to tissue shrinkage or magnified by thickness of the amber. Fossils rarely have the exact same morphological features as related extant lineages. Nor can extant lineages have features of extinct lineages that existed some 20-30 MYA. Aside from this, it is obvious that only a fraction of the seeds of extant orchids are known, not to mention the seeds of extinct orchids. So stating that this is not an orchid seed or the fungus is not a symbiont is meaningless. *Mycophoris elongatus* is available for study at Oregon State University. Thus far, no one has shown any interest in examining it nor has anyone shown any extant dust seeds from other plant families that have features of *Mycophoris elongatus*.

Indirect Evidence of Orchids in Amber

Orchid Bees in Dominican Amber

New World orchid bees have attracted the attention of entomologists and botanists for their pollination activities, often being the sole pollinators of tropical orchid lineages (Poinar 1998A). These bees are normally fairly large and often brilliantly colored with metallic markings. A pair of female orchid bees in Dominican amber shared characters with three of the five known New World genera of orchid bees. *Paleoeuglossa melissiflora* Poinar (1998B), the first described orchid bee in amber, possessed a metallic sheen similar to members of the extant genera *Euglossa* and *Eufresea* and a scutellar tuft found in extant *Euglossa* and *Eulaema*. However, its protruding clypeus and genal projections (Figure 24) were unique.

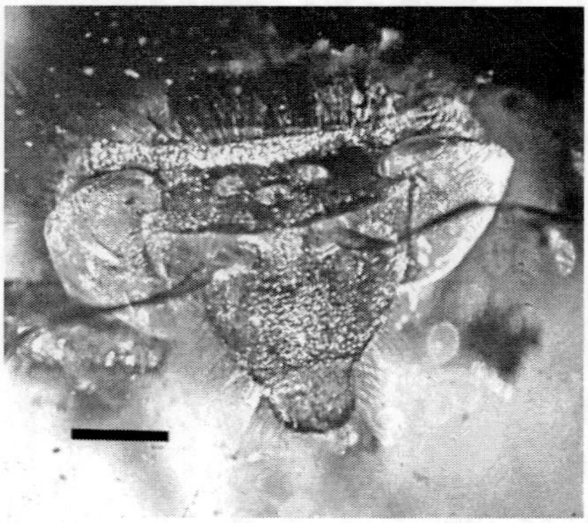

Figure 24. Head of the orchid bee, *Paleoeuglossa melissiflora* in Dominican amber. Scale bar = 1.0 mm.

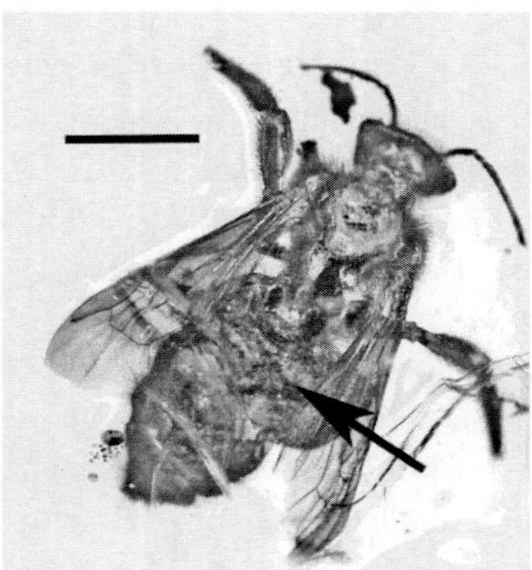

Figure 25. A Dominican amber orchid bee with what could be one or more pollinia attached to its abdomen (arrow). Scale bar = 4.0 mm.

Finding a pair together in amber suggests that they were collecting resin for nest construction when entrapped. A second Dominican amber orchid bee was described as *Euglossa moronei* (Engel, 1999) and bears synapomorphic characters that place it unambiguously in the extant genus *Euglossa* Latreille, 1802.

An undescribed orchid bee in Dominican amber carries a structure on its back that could be one or more pollinaria (Figure 25). Orchid bees have since been extirpated from Hispaniola (Poinar 1998B, 2013).

References

Ackerman, J. D., and Michael R. Mesler. "Pollination Biology of *Listera cordata* (Orchidaceae)." *American Journal of Botany* 66, no. 7 (1979): 820-24.

Arditti, J. and A. Ghani. "Tansley Review No. 110: Numerical and physical properties of orchid seeds and their biological implications." *New Phytologist* 145 (2000): 367-421.

Arditti, J. "Factors Affecting the Germination of Orchid Seeds." *Botanical Review* 33, no. 1 (1967): 1-97.

Bauch, R. "Untersuchungen über zweisporige Hymenomyceten I. Haploide parthenogenesis bei *Camarophyllus virgineus*." *Zeitschrift für Botanik* 18 (1926): 337-87. ["Studies on two-pore Hymenomycetes I. Haploid parthenogenesis in Camarophyllus virgineus." *Journal of Botany*]

Berggren, W. A., John A. Van Couvering. "The Late Neogene: Biostratigraphy, Geochronology and Paleoclimatology of the Last 15 Million Years in Marine and Continental Sequences." *Palaeogeography, Palaeoclimatology, Palaeoecology* 16, no. 1-2 (1974): 1-216.

Conran, J. G., Jennifer M. Bannister, and Daphne E. Lee. "Earliest Orchid Macrofossils: Early Miocene *Dendrobium* and *Earina* (Orchidaceae: Epidendroideae) from New Zealand." *American Journal of Botany* 96, no. 2 (2009): 466-74.

Czeczott, H. "The flora of the Baltic amber and its age." *Prace Muzeum Ziemi (Paleobotaniczne) Warszawa* 4 (1961): 119-145.

Draper, G., Paul Mann, and John Lewis, F. "Hispaniola." In *Caribbean geology: an introduction*, edited by Stephen Donovan, K., and Trevor Jackson, A., 129-150, Kingston, Jamaica: The University of the West Indies Publishers' Association, 1994.

Dressler, R. L. *Phylogeny and Classification of the Orchid Family*. Portland: Dioscorides Press, 1993.

Duque-Buitrago, C. A., Néstor Fabio Alzate-Quintero, and J. Tupac Otera. "Nocturnal pollination by fungus gnats of the Colombian endemic species, *Pleurothallis marthae* (Orchidaceae: Pleurothallidinae)." *Lankesteriana* 13 (2014): 407-17.

Endress, P. K. *Diversity and evolutionary biology of tropical flowers*. Cambridge: Cambridge University Press, 1994.

Engel, M. S. "The First Fossil *Euglossa* and Phylogeny of Orchid Bees (Hymenoptera, Apidae, Euglossini). *American Museum Novitates* 3272 (1999): 1-14.

Givnish, T. J., Daniel Spalink, Mercedes Ames, Stephanie P. Lyon, Steven J. Hunter, Alejandro Zuluaga, William J. D. Iles, et al., "Orchid Phylogenomics and Multiple Drivers of Their Extraordinary Diversification." *Proceedings. Biological Sciences* 282, no. 1814 (2015): 20151553.

Gołaszewska, E., Joanna Gadziszewska, and Małgorzata Latałowa. "First Record of Orchid Subfossil Seeds – The Abundant Occurrence of *Epipactis palustris* (L.) Crantz and *Dactylorhiza* spp. Seeds in Early Holocene Sediments from Central Europe." *Review of Palaeobotany and Palynology* 265 (2019): 1-12.

Halbwachs, H., and Claus Bässler. "Gone with the Wind – a Review on Basidiospores of Lamellate Agarics." *Mycosphere* 6, no. 1 (2015): 78-112.

Iturralde-Vinent, M. A., and R. D. E. MacPhee. "Age and Paleogeographical Origin of Dominican Amber." *Science* (New York, N.Y.) 273, no. 5283 (1996): 1850-52.

Kalshoven, L. G. E., and P. A. van der Laan. *Pests of Crops in Indonesia*. Jakarta, Indonesia: P. T. Ichtiar Baru Press, 1981.

Lambert, J. B., J. S. Frye, and G. O. Poinar. "Amber from the Dominican Republic: Analysis by Nuclear Magnetic Resonance Spectroscopy." *Archaeometry* 27, no. 1 (1985): 43-51.

Larsson, S. G. *Baltic amber – a palaeobiological study*. Klampenborg: Scandinavian Science Press, 1978.

Mesler, M. R., James D. Ackerman, and Karen L. Lu. "The Effectiveness of Fungus Gnats as Pollinators." *American Journal of Botany* 67, no. 4 (1980): 564-67.

Poinar, G. O., Jr. "*Hymenaea protera* sp.n. (Leguminosae, Caesalpinioideae) from Dominican Amber Has African Affinities." *Experientia* 47, no. 10 (1991): 1075-82.

Poinar, G. O. Jr. "Orchid bees." *Orchids* 67 (1998A): 578.

Poinar, G. O., Jr, and Roberta Hess. "Preservative Qualities of Recent and Fossil Resins: Electron Micrograph Studies on Tissue Preserved in Baltic Amber." *Journal of Baltic Studies* 16, no. 3 (1985): 222-30.

Poinar, G., Jr, and Alex E. Brown. "*Hymenaea mexicana* sp. nov. (Leguminosae: Caesalpinioideae) from Mexican Amber Indicates Old World Connections." *Botanical Journal of the Linnean Society* 139, no. 2 (2002): 125-32.

Poinar, G., Jr, and Finn N. Rasmussen. "Orchids from the Past, with a New Species in Baltic Amber." *Botanical Journal of the Linnean Society* 183, no. 3 (2017): 327-33.

Poinar, G., Jr. "Beetles with Orchid Pollinaria in Dominican and Mexican Amber." *American Entomologist* 62, no. 3 (2016B): 172-77.

Poinar, G., Jr. "Desiomorphs in Amber." *American Entomologist* 58, no. 4 (2013): 214-23.

Poinar, G., Jr. "Developmental Stages of the Fungus, *Synaptomitus orchiphilus*, in the Germinating Seed, *Mycophoris elongatus* (Orchidaceae), in Dominican Amber," *Historical Biology* (2017B): https://doi.org/10.1080/08912963.2017.1411352.

Poinar, G., Jr. "Two New Genera, *Mycophoris* gen. nov., (Orchidaceae) and *Synaptomitus* gen. nov. (Basidiomycota) Based on a Fossil Seed with Developing Embryo and Associated Fungus in Dominican Amber," *Botany* 95, no. 1 (2017A): 1-8.

Poinar, G., Jr., and Roberta Poinar. *The Amber Forest.* Princeton: Princeton University Press, 1999.

Poinar, G. "Orchid Pollinaria (Orchidaceae) attached to stingless bees (Hymenoptera: Apidae) in Dominican Amber." *Neues Jahrbuch Für Geologie Und Palaontologie* 279, no. 3 (2016A): 287-93.

Poinar, G. "*Paleoeuglossa melissiflora* gen. n., sp. n. (Euglossinae: Apidae), Fossil Orchid Bees in Dominican Amber." *Journal of the Kansas Entomological Society* 71, no. 1 (1998B): 29-34.

Poinar, Jr., G. O., and M. Mastalerz. Taphonomy of fossilized resins: determining the biostratinomy of amber. *Acta Geologica Hispanica* 35, no. 1–2 (2000): 171-82.

Ramírez, S. R., Barbara Gravendeel, Rodrigo B. Singer, Charles R. Marshall, and Naomi E. Pierce. "Dating the Origin of the Orchidaceae from a Fossil Orchid with Its Pollinator." *Nature* 448, no. 7157 (2007): 1042-45.

Ridley, H. N. "Orchideae and Apostasiaceae of the Malay Peninsula." *Botanical Journal of the Linnean Society* 32 (1896): 213-416.

Schlee, D. "Das Bernstein-Kabinett." *Stuttgarter Beiträge für Naturkunde Ser C.* 28 (1990): 1-100. [The Amber Cabinet. *Stuttgart contributions to natural history Ser C.*]

Schmid, R., and M. J. Schmid. "Fossil history of the Orchidaceae." In *Orchid biology reviews and perspectives*, edited by Joseph Arditti, 25-45. Ithaca: Cornell University Press, 1977.

Selosse, M. A., Mark Brundrett, John Dearnaley, Vincent S. F. T. Merckx, Finn Rasmussen, Lawrence W. Zettler, and Hanne N. Rasmussen. "Why *Mycophoris* is not an Orchid Seedling, and why *Synaptomitus* is not a fungal symbiont within this fossil." *Botany* 95, no. 9 (2017): 865-68.

Singer, R. B., Barbara Gravendeel, Hugh Cross, and Santiago R. Ramirez. "The Use of Orchid Pollinia or Pollinaria for Taxonomic Identification." *Selbyana* 29, no. 1 (2008): 6-19.

Stpiczyńska, M., Bartosz J. Płachno, and Kevin L. Davies. "Nectar and Oleiferous Trichomes as Floral Attractants in *Bulbophyllum saltatorium* Lindl. (Orchidaceae)." *Protoplasma* 255, no. 2 (2018): 565-74.

van der Cingel, N. A. *An atlas of orchid pollination: America, Africa, Asia and Australia*. Rotterdam: A. A. Balkema, 2001.

Wolfe, A. P., Ryan C. McKellar, Ralf Tappert, Rana N. S. Sodhi, and Karlis Muehlenbachs. "Bitterfeld Amber Is Not Baltic Amber: Three Geochemical Tests and Further Constraints on the Botanical Affinities of Succinite." *Review of Palaeobotany and Palynology* 225 (2016): 21-32.

Zimmerman, E. C. *Australian Weevils (Coleoptera: Curculionoidea).* Vol. 6. East Melbourne, Victoria: CSIRO, 1992.

In: Orchidaceae
Editor: Vladan Djordjević

ISBN: 978-1-68507-189-9

Chapter 2

How the Number of Orchid Populations Quickly Declined in the Czech Republic During the Last 150 Years? The Case of Less Threatened Species

Zuzana Štípková[1,2,*] and Pavel Kindlmann[1,2]
[1]Global Change Research Institute,
Academy of Sciences of the Czech Republic, Czech Republic
[2]Institute for Environmental Studies, Faculty of Science,
Charles University, Czech Republic

Abstract

In the contemporary period of massive extinction of species it is especially important to understand the way in which species distribution patterns and abundances change over time. This may be affected by the political regime. In Czechoslovakia, before 1948, the traditional

* Corresponding Author's E-mail: zaza.zuza@seznam.cz.

management practice was fields and meadows that were mown or grazed, with low intensity of agriculture based on small fields and low application of fertilizers. After 1948, the practices in Czechoslovakia changed dramatically, with small fields being merged into huge fields and subsidies provided for fertilizers, so that the amount of nutrients in the soil increased rapidly and negatively affected many rare species. After the change in the regime in 1989, subsidies for fertilizers stopped, which caused a great decline in their use and subsequent increase in the distribution of many rare species.

In a previous study, Štípková and Kindlmann (2021b) analysed, for different species of orchids, the temporal trends in the numbers of sites, where living populations of the species studied were reported at the time considered. They used a dataset of orchid records of the Nature Conservation Agency of the Czech Republic, which covers a period of about 150 years and includes more than 115 000 records.

Contrary to these analyses, which were done for rare species of Czech orchids, i.e., for threat categories A1 and C1, based on the Red List classification for the Czech Republic, here we present results for the less endangered species: for threat categories C2-C4.

We conclude that the numbers of sites, where living populations were reported, declined less in species with many historically recorded sites, maybe because these species are widely distributed and not very demanding in terms of site quality. This decline was largest during the communist era (1948-1989), compared with other periods. In the case of species with few historically recorded sites this decline was much greater, maybe because of their special requirements on site quality, but not dependent on the time period.

This indicates that quality of the sites of extinction-prone species should be maintained, which is not so important in the case of the more abundant species, as their extinction is less likely. The fact that these losses occurred mainly during the communist era may indicate that the large disturbances occurring then (conversion into building sites, large inputs of pesticides or artificial fertilizers applied to adjacent fields, etc.) were the cause.

Keywords: conservation, loss of populations, number of sites, orchids, protection of natural habitats

INTRODUCTION

The contemporary massive extinction of species (IPBES 2019) makes it especially important to understand the way, in which patterns in the distributions of species and their abundances change over time (Gaston and Blackburn 2000). The main reasons, why the abundance and distributions of many plant and animal species have declined dramatically in recent decades, are habitat loss and fragmentation (Fischer and Stöcklin 1997; Kull et al. 2002). During the recent centuries, rapid industrial development resulted in a highly fragmented and degraded landscape (Bilz et al. 2011). Changes in farming and demographic trends over recent decades brought about dramatic changes in the composition of land cover in Europe (Henle et al. 2008). In summary, the most important causes of habitat destruction in many parts of Europe are urbanization, changes in land use and intensification of agriculture (Stewart 1992; Tsiftsis et al. 2011; Kull et al. 2016).

In the Czech Republic (western part of former Czechoslovakia, which consisted of Bohemia, also named Czechia, and Slovakia, until the splitting of the country in 1993), land use has undergone very marked changes, which substantially differed from those in Western Europe, because of the different political regimes in Central and Western Europe. Before 1948, the traditional management of fields and meadows consisted of mowing and grazing with generally low intensity agriculture of small fields and low application of fertilizers. After 1948, the practices in Czechoslovakia changed, with small fields being merged into huge fields and cost of fertilizers was subsidized, which resulted in a dramatic increase in their use. The result of these changes in management was that the abundances of rare species in many of these areas decreased (Adams and Adams 1971). Following the change in regime in 1989, subsidies for fertilizers stopped and consequently there was a great decline in the use of fertilizers (Adams and Adams 1971; Wädekin 1982) and subsequent increase in the abundances of many rare species.

Here we use orchids as the model group, because they are one of the most threatened groups of species worldwide. With approximately 28,500 species (Govaerts 2020), Orchidaceae is one of the most diverse and widespread family of flowering plants (Swarts and Dixon 2009). The reasons why most species of orchids are threatened or endangered (Cribb et al. 2003) and disappearing worldwide (Kull and Hutchings 2006) is habitat loss, eutrophication and fragmentation (Wotavová et al. 2004; Janečková et al. 2006; Kull and Hutchings 2006; Kull et al. 2016). Their usually low pollination success and low levels of fruit set are also linked to their rarity (Darwin 1862; Neiland and Wilcock 1998). Due to the activities of professionals, as well as amateurs, orchids are well recorded and studied in many countries in Europe (Kull et al. 2006). The availability of detailed records, especially for different time periods, provide opportunities for comparative analyses of declines in species over time.

A good knowledge of environmental factors that are correlated with decline in abundance is very important for making proper recommendations of how to manage sites, where orchids still survive and prevent them from becoming unsuitable (Kull and Hutchings 2006).There isa large number of studies dealing with orchids, but the critical information necessary for their conservation is still missing.

An important category of orchids that requires special conservation measures is that of the orchids classified in the most threatened category according either to the IUCN guidelines or other national classifications, e.g., the classification that is used in the Czech Republic (Danihelka et al. 2012; Grulich 2012). These lists of species provide essential information on trends in the loss of biodiversity (Hoffmann et al. 2008; Zamin et al. 2010). Red list assessments might differ among countries (Maes et al. 2015), because these are not always based on the latest IUCN red list categories and criteria (Sharrock and Jones 2009). However, some general patterns at the landscape level might be found using national red list data (Zamin et al. 2010).

In a recent study, Štípková and Kindlmann (2021b) analysed the temporal trends in changes in the numbers of living populations – those, where living populations of the species were reported at the time considered – of the most endangered species of orchids, i.e., for threat categories A1 and C1, based on the Red List classification for the Czech Republic and species that went extinct during the last 150 years. The dataset of orchid records they used was based on the database of the Nature Conservation Agency of the Czech Republic. They put their findings into the context of relevant changes in agriculture during this period.

Štípková and Kindlmann (2021b) found that the vast majority of orchids disappeared from many of their historical localities in all time intervals analysed. The numbers of living populations declined by 8–92%, depending on the species. The most threatened orchid species in the Czech Republic were *Spiranthes spiralis*, *Anacamptis palustris*, *Epipogium aphyllum* and *Goodyera repens*. Štípková and Kindlmann (2021a,b) conclude that these declines all seem to be closely related to changes in agricultural practices in open as well as in forest habitats. They also stress that preserving suitable orchid habitats seems to be the key for keeping the Czech orchid flora.

There is one question that remains: do the patterns observed by Štípková and Kindlmann (2021b) for the most endangered species of orchids in the Czech Republic also hold for the remaining ones? Štípková and Kindlmann (2021b) found that there were almost no differences in the rate of decline of the numbers of living populations between the period when the communist regime governed in Czechoslovakia (1948-1989) and the subsequent period of free market economy (1989-2020). This is contrary to expectation, as the application of fertilizers, at least on-nitrogen fertilizers, declined after 1989 (Štípková and Kindlmann 2021b; Štípková et al. 2021), which in theory should favour orchids. In addition, the post-communist period is considered to be more nature friendly than the communist one. This raises the question: why is this not reflected in the percentage loss of the numbers of orchid living populations? Maybe this can only be observed for the less endangered orchid species.

It is also evident that much more attention was paid to nature after the fall of communism. There are now governmental agencies involved in nature protection, organising the management of orchid sites and also the number of orchid enthusiasts has increased dramatically. Therefore, it is reasonable to believe that most of the currently living populations of the rarest species were known even before 1989, as they were seen as important, while some less important sites of abundant species were ignored before 1989, when far fewer people looked for orchids and no computer databases were available for storing large amounts of data. If this is true, then one might expect an increase in the number of living populations for abundant species of orchids, not because these orchids have become more widely spread, but because previously ignored sites are now newly included in the databases.

For all the above reasons, we decided to repeat the analyses we did previously for rare species of Czech orchids (for threat categories A1 and C1, based on the Red List classification for the Czech Republic) in Štípková and Kindlmann (2021b) now for the land use has undergone very less endangered species of orchids (for threat categories C2-C4). Hence, we analyse the trends in the changes in the numbers of populations
(= living populations) of less threatened, and therefore usually more abundant species of orchids and put the available data into the context of the changes that occurred in agricultural practices in Czechoslovakia/ Czech Republic.

Materials and Methods

Study Site

The Czech Republic is in central Europe and its flora is very well studied. Its average altitude is 450 m a.s.l., according to the calculation based on the 30-sec altitude layer (Fick and Hijmans 2017) which was performed in ArcMap 10.1 (ESRI 2012). The country is covered mainly by highlands of moderate altitude, whereas higher mountains occur at the borders with other countries (mainly in the north and south). The climate of the Czech Republic is typically temperate with cold, cloudy winters and hot summers. However, there are some regional and local differences due to the relief that forms the complex topography in this area (Palacký University Olomouc 2020). Because the Czech Republic is a relatively small country in terms of latitudinal range, temperature and precipitation are mostly affected by local heterogeneity and altitude (Štípková et al. 2020).

The Database

The dataset of orchid records we used is based on that of the Nature Conservation Agency of the Czech Republic. The records cover a period of about 150 years and include all the known literature sources and reports from people of records of orchids, the identities of which are subsequently checked by specialists. The database includes more than 115,000 individual records.

Each site is characterized by its GPS coordinates. Because of the enormous number of records and extremely long timespan covered, the cross-checking for duplicate records was not consistently done, especially in the older data. This means that in many cases, the same site may be called by different names and the GPS coordinates may differ for the same sites in different records in the database.

We resolved this problem by using ArcMap 10.1 (ESRI 2012) to create a grid of 1×1 km squares. For each square, we then considered any recording, the GPS coordinates of which fall within this square, as the same site (population). This is reasonable, as typical orchid sites cover several acres or hectares, and small adjacent sites are likely to be sub-sites of a larger site, or, in other words, an orchid population.

For each of the sites (squares in our grid where an orchid population was recorded), we determined the latest year when the orchid was still observed at the site. If this year was 1990 or later, the site was considered to be still occupied (except of *Dactylorhiza curvifolia* and *Gymnadenia odoratissima*, which are commonly considered to be extinct in the Czech Republic, despite of being found during 1990-2020). If the last record of living population at a site was prior 1990, this date was considered to be the date of extinction of the orchid population at this site.

It is unlikely that this resulted in a bias in our calculations. First, the Nature Conservation Agency of the Czech Republic meticulously rechecked all the sites, where the orchids were likely to survive. Second, if the date of the last record was prior 1990, then of course the population might have survived for several additional years after this observation. However, as the dates of extinction are categorized into intervals of approximately 20 years duration (Table 1), it is most likely that the actual date of extinction is in the same period of 20 years as the date of the last record.

Nomenclature

Classification, nomenclature and Red List classification (threat categories) of native taxa is based on Grulich (2012), Grulich and Chobot (2017) and Danihelka et al. (2012).

Data Analysis

We categorized a total of 64 species of orchids into threat categories based on the Red List classification for the Czech Republic (Danihelka et al. 2012; Grulich 2012; Grulich and Chobot 2017). For the purpose of this study, we analysed only the 29 species that are in the category of highly endangered species (C2), endangered species (C3) and less threatened species (C4). We categorized the C2, C3 and C4 species according to the number of sites recorded in the database: (a) C2 species recorded at less than 100 sites, (b) C2 species recorded at 101-300 sites, (c) C2 species recorded at more than 300 sites, (d) C3 species recorded at less than 1000 sites, (e) C3 species recorded at 1000 and more sites and (f) C4 species. Within each of these categories, we ranked the species according to the percentage of sites, at which they survived up to the present time, from the lowest to the largest. For each species, we calculated the number of sites at which extinction was recorded during the selected time periods (see Table 1).

Thus, we mainly considered political events that are most likely to have affected the agricultural practices, as milestones, rather than what was recorded at a particular interval of time. We looked for patterns in when orchids disappeared from sites for the different species and attempted to associate them with changes in the society and political regimesthat resulted in severe changes in agricultural practices.

Table 1. Explanation of the selected periods

Time period	Explanation
Before 1900	very old data from the 19th century
1901-1918	prior to the end of the 1st World War and founding of Czechoslovakia
1919-1938	during the existence of the "1st" Czechoslovak Republic
1939-1948	2nd World War and the period before the communist putsch in 1948
1949-1968	Czechoslovakia under the communist regime, before the Prague Spring
1969-1989	the period after the Soviet invasion and before the Velvet Revolution
1990-2020	the period after the Velvet Revolution – free market economy in Czechoslovakia

The results we present in the figures for each interval are the numbers of sites, for which the latest year when the orchid was still found to be present at the site was in that particular interval. They are referred to as “newly unoccupied sites.” The reason for this notation is that – as we described above – if a site was never recorded as occupied later than in this interval, the population here probably went extinct during this interval. The numbers of sites recorded as occupied (living populations by our definition) in the last interval are considered to be currently occupied.

RESULTS AND DISCUSSION

The total numbers of sites for each species that where included in our database, i.e., which were recorded over the last 150 years, together with the percentages of these that remained occupied up to the present time is shown in Table 2.

Table 2. The total number of sites and their proportion occupied at the last interval for each orchid species analysed in the Czech Republic

Species	**Total number of sites**	**Sites occupied (%)**	**Species**	**Total number of sites**	**Sites occupied (%)**
Cephalanthera damasonium	2002	89	*Epipactis muelleri*	98	73
Cephalanthera longifolia	919	88	*Epipactis palustris*	620	76
Cephalanthera rubra	355	71	*Epipactis purpurata*	638	83
Corallorhiza trifida	196	55	*Epipactis voethii*	9	100
Cypripedium calceolus	314	59	*Gymnadenia conopsea*	1271	74
Dactylorhiza fuchsii	2716	90	*Neottia nidus-avis*	2913	86
Dactylorhiza majalis	6943	86	*Neottia ovata*	3245	84
Dactylorhiza sambucina	541	76	*Orchis militaris*	419	70
Dactylorhiza viridis	179	38	*Orchis pallens*	299	83
Epipactis albensis	202	58	*Orchis purpurea*	275	70
Epipactis atrorubens	350	76	*Platanthera bifolia*	3771	81
Epipactis greuteri	84	93	*Platanthera chlorantha*	1140	89
Epipactis helleborine	4954	89	*Pseudorchis albida*	79	37
Epipactis leutei	5	100	*Traunsteinera globosa*	358	75
Epipactis microphylla	168	42			

Species are sorted alphabetically.

In the following subsections, the results are individually analysed for individual groups of orchids in each of the time intervals outlined in the section Data Analysis.

C2 Species with Less Than 100 Sites

Figure 1 shows the results for the seven highly endangered (C2) species that were found historically at less than 100 sites. It seems that one of the most endangered species included in this study is *Pseudorchis albida* with only 29 living populations in the last time period (Figure 1). Together with *Dactylorhiza viridis*, they represent the largest rate of decline of the numbers of living populations among all studied species, 37% and 38%, respectively. It is not a big surprise, as both species typically occur in meadows that were affected by changes in agricultural practices in the past. *Pseudorchis albida* has suffered a significant decline in distribution not only in the Czech Republic, but also in Britain and other European countries. In Britain, it has been lost from more than 65% of its historical sites (Jersáková et al. 2011). A threat to *P. albida* is mainly connected with its relatively high rarity and overgrowing of its localities. Populations of this species tend to be small in the Czech Republic, so the species may be overlooked because of its inconspicuousness. Most historical populations (living populations) of *D. viridis* in lowlands are extinct now and the major threat to this species is destruction of its natural habitats (Průša 2005). Only *Epipactis leutei* and *E. voethii* are now present at less living populations than *P. albida*, but the situation in the *Epipactis* group is very complicated. The sites in the Czech Republic of the two lately mentioned species of *Epipactis* were discovered in the last studied period – *E. leutei* in 2005 and *E. voethii* in 2001 (Průša 2005). According to their recent discovery for the Czech flora, it is very probable that the numbers of their known localities will increase in the future.

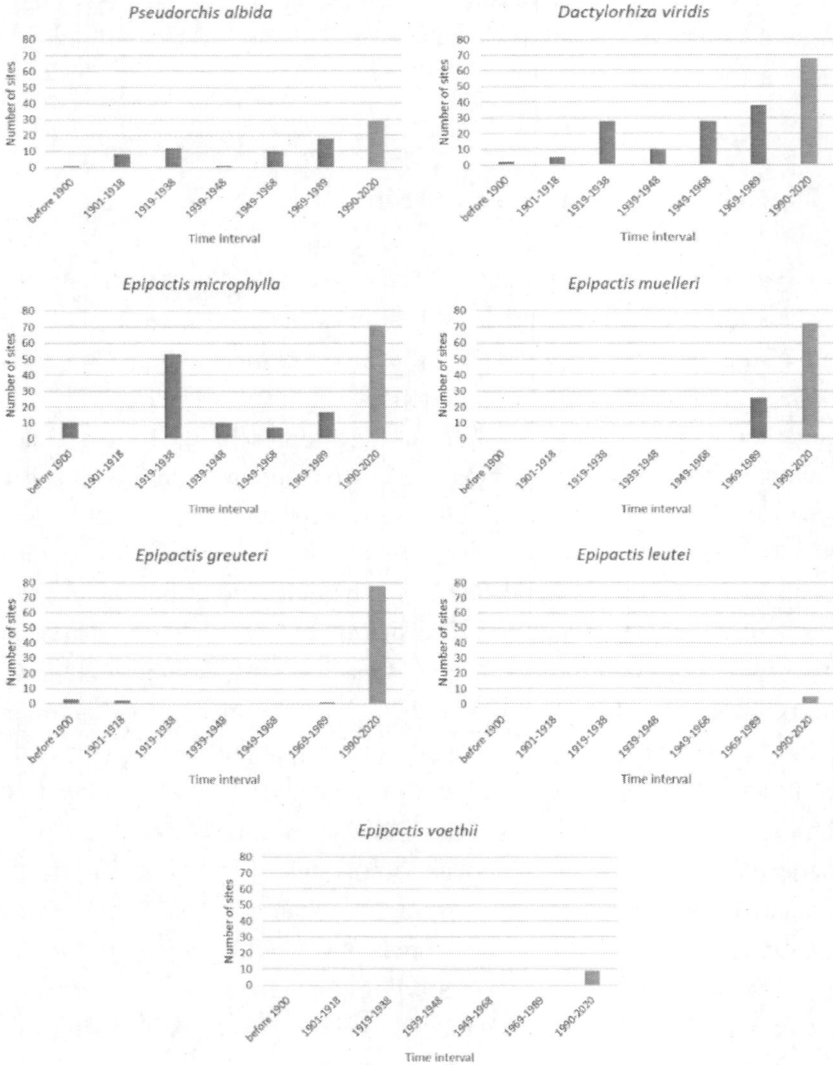

Figure 1. Numbers of sites of the seven species in the C2 category (highly endangered species) that were found historically at less than 100 sites for the six studied periods: blue columns are the numbers of sites that were occupied by orchids for the last time during the given interval and then not anymore, so the height of the column indicates the rate of decline of the numbers of living populations in the particular interval. The green bars indicate the numbers of living populations in the last interval, which we consider as living currently.

The vast majority of *Epipactis* species typically occur in various kinds of forests that were much less affected by the changes in the agricultural and management practices (Skaloš et al. 2012). The general pattern is that the number of currently living populations is always, for all species considered, larger than the number of sites, in which the local populations went extinct during any of the 6 intervals depicted, prior 1990 (Figure 1).

Considerable numbers of sites, in which the local populations went extinct, was observed for *Pseudorchis albida*, *Dactylorhiza viridis* and *Epipactis microphylla* prior 1990, while almost all sites remained occupied after 1990 for the remaining four *Epipactis* species because of the reason mentioned above.

In the course of time, i.e., from the 1st (before 1900) to the 6th (1969-1989) time interval, there was a tendency for an increase of the numbers of sites, in which the local populations went extinct per interval (we define it as the rate of decline or rate of extinction of the numbers of living populations – Figure 1). This was especially conspicuous in the case of *Dactylorhiza viridis* and partially also for *Pseudorchis albida* (Figure 1): the largest numbers of local populations went extinct during the intervals 1949 to 1968, and especially 1969 to 1989, i.e., during the communist regime. There is no clear temporal tendency for changes in the numbers of living populations for any of the *Epipactis* species because of the reason mentioned above.

C2 Species with 100 to 300 Sites

Figure 2 shows the results for the eight highly endangered (C2) species that were found historically at 100 to 300 sites. Again, there was a tendency for an increase of the rate of decline of the numbers of living populations in the course of time, with this increase being most conspicuous in the period 1969-1989.

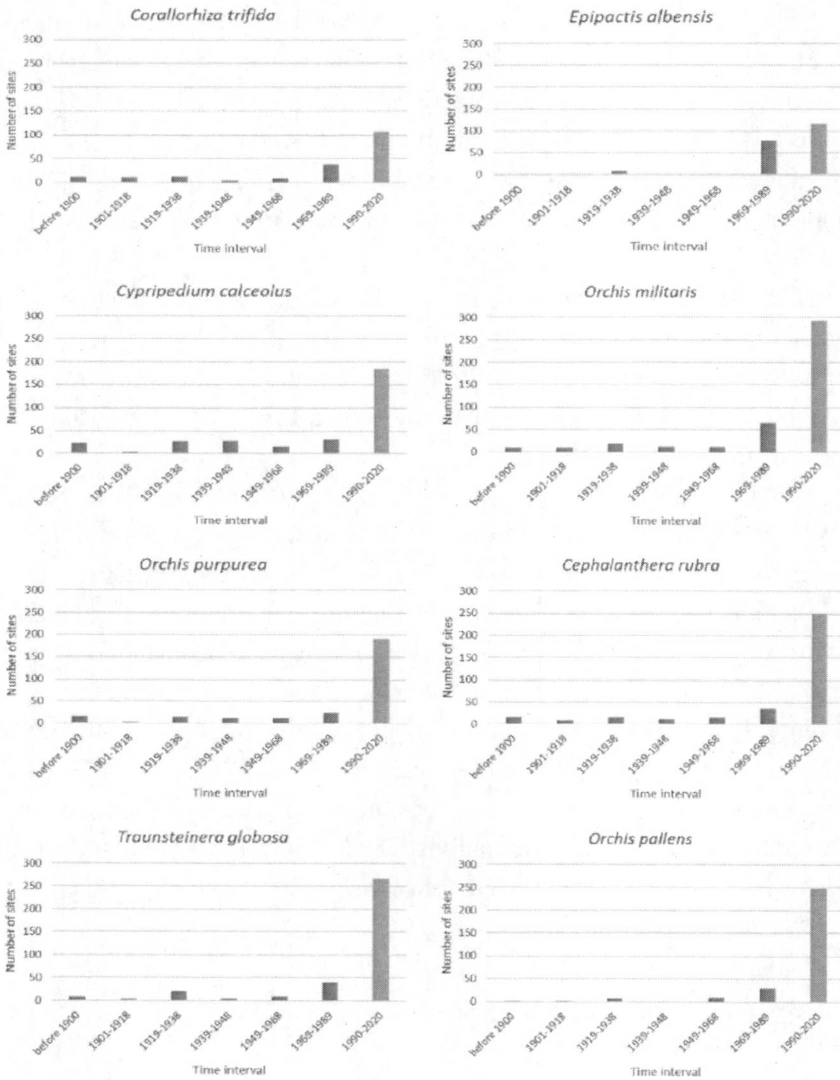

Figure 2. Numbers of sites of the eight species in the C2 category (highly endangered species) that were found historically at 100 to 300 sites for the six studied periods: blue columns are the numbers of sites that were occupied by orchids for the last time during the given interval and then not anymore. The green bars indicate the numbers of living populations in the last interval, which we consider as living currently.

However, the number of living populations in the present time (i.e., with positive records during the time interval 1990 to 2020), is much higher when compared with any of the previous time intervals. This dominance of the present numbers of living populations is much more conspicuous than in the case of the species that were found historically at less than 100 sites (Figure 1).

The majority of species from this group declined because of destruction of their natural habitats, succession and changes in management practices in the past. The decline in distributional range of *Corallorhiza trifida* was also reported in Estonia and United Kingdom (Kull and Hutchings 2006) and this species is critically endangered in Romania (Molnár 2011). *Corallorhiza trifida* is fully dependent on mycorrhiza and therefore it is limited by any intervention into habitat quality that may affect composition of soil microflora (Průša 2005). The distribution of *Traunsteinera globosa*in the Czech Republic has changed in the past. It was considered extinct in Czechia, the western part of the Czech Republic, but it is still living in protected areas in the eastern, Moravian part of the country, where a proper management is provided (Průša 2005). The prognosis of the future fate of *Epipactis albensis* is much more optimistic. It was firstly described in the Czech Republic, later it was found in Slovakia and in other parts of Europe (Průša 2005; Molnár and Sramkó 2012; Ljubka et al. 2014). This demonstrates that its distribution area is much larger. Only three (*Orchis militaris*, *O. purpurea* and *Traunsteinera globosa*) out of the eight species in this group occur mostly in open habitats or shrubs while the remaining ones thrive in forests.

C2 Species with More than 300 Sites

There are three species that were historically found at more than 300 sites, in the highly endangered (C2) group (Figure 3). For these species, the percentages of currently occupied sites are quite high (76%, 76% and 74% respectively), i.e., much higher than in the case of C2 species with 100 to 300 sites. The rate of decline of the numbers of living populations slightly increased in time, similarly to the case of C2 species with 100 to

300 sites, the largest increase being againin the period 1969-1989 (Figure 3).

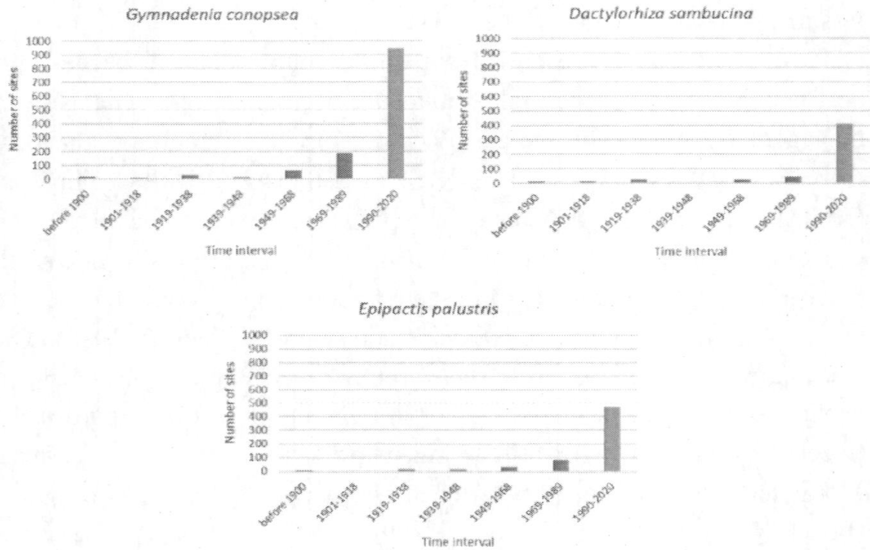

Figure 3. Numbers of sites of the three species in the C2 category (highly endangered species) that were found historically at more than 300 sites for the six studied periods: blue columns are the numbers of sites that were occupied by orchids for the last time during the given interval and then not anymore, so the height of the column indicates the rate of decline of the numbers of living populations in the particular interval. The green bars indicate the numbers of living populations in the last interval, which we consider as living currently.

The main reason for the decline of all species in this group is destruction of their natural habitats and changes in management regimes connected mainly with absence of traditional management (Průša 2005; Meekers et al. 2012; Jacquemyn et al. 2014; Kaplan et al. 2017) as these species typically occur in open habitats or shrubs. In the Czech Republic, the highest declines occurred between 1969 and 1989, when their population numbers became weaker and now they are often absent in lowlands (Průša 2005). The numbers of populations and sizes of populations in Britain have declined markedly during the recent decades (Meekers et al. 2012). Conservation of *G. conopsea* populations requires appropriate management regimes, like mowing or grazing (Průša 2005;

Meekers et al. 2012). *Dactylorhiza sambucina* became extinct at a significant portion of its historical localities in the second half of the 20th century (Průša 2005; Kaplan et al. 2017). On the other hand, many of its recent localities are under protection and appropriate management is applied (Průša 2005). The last species in the group, *Epipactis palustris*, has suffered a strong decline in its distribution range also in Britain, with most losses of local populations occurring before 1930 (Jacquemyn et al.2014), contrary to the Czech Republic, where the highest decline of the numbers of living populations occurred between 1969 and 1989. The highest decline of living populations in the second half of the 20th century was also reported for another marsh orchid, *Dactylorhiza majalis*, from the C4 group that sometimes co-occurs with *E. palustris* at its localities.

C3 Species with Less than 1000 Sites

Figure 4 presents the situation of the three species that belong to the group of the endangered orchids and were historically recorded in less than 1000 sites. The survival of populations of these species at these sites is higher than 75% (the highest for *Cephalanthera longifolia*, 88%). Figure 4 shows that the tendency for the rate of decline of the numbers of living populations slightly increased in time with the highest extinction rate in the period 1969-1989, similarly to the previous two cases – those of C2 species with 100-300 sites and those with more than 300 sites. There were almost no extinctions of populations prior 1948 in all three species and even the numbers of their sites, where the local populations went extinct during the communist era, were very small (Figure 4).

For all the species, occurring mainly in forests, destruction of their natural habitats is the main threat in the Czech Republic, similarly as in other groups of orchids in this study. The distribution of *Epipactis atrorubens*is dependent on calcareous substrates in general, in the Czech Republic it is reported to be able to spread also with calcareous gravel used for road constructions (Průša 2005). A damage caused by grazing animals is mentioned as an additional factor for this species decline

(Walker 2015). In Serbia, *Epipactis purpurata* is listed as vulnerable species (Djordjević et al. 2017). On the other hand, in the Czech Republic, part of its localities lies within protected areas, which may help to preserve the species for the future (Průša 2005). The decline in the numbers of living populations of *Cephalanthera longifolia* was low in the Czech Republic and a similar trend was reported also from Estonia, contrary to the data from the United Kingdom, where the decline was much higher (Kull and Hutchings 2006).

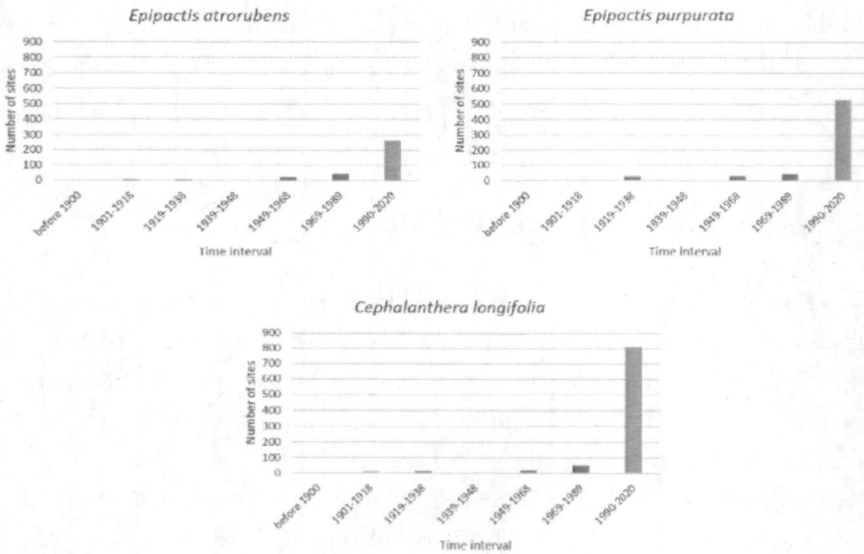

Figure 4. Numbers of sites of the three species in the C3 category (endangered species) that were found historically at less than 1000 sites for the six studied periods: blue columns are the numbers of sites that were occupied by orchids for the last time during the given interval and then not anymore, so the height of the column indicates the rate of decline of the numbers of living populations in the particular interval. The green bars indicate the numbers of living populations in the last interval, which we consider as living currently.

C3 Species with 1000 and More Sites

There are three species with 1000 and more sites among the endangered group (C3). The rate of extinction of their local populations at the sites was very low compared to total number of their sites recorded in the database (Figure 5). Their percentage survival (number of currently living populations over total historically recorded number of sites) reaches nearly 90%. In the case of *Platanthera chlorantha*, the numbers of sites, where the local populations went extinct, are too low to be visible in Figure 5 (42 sites, where the local population went extinct, as the highest in the period 1969-1989). *Dactylorhiza majalis* is the most abundant orchid species mentioned in the database with the current number of its living populations reaching almost 6000.

The tendency for the numbers of extinctions of local populations of these species to grow in time with the highest extinction rate being in the period 1969-1989, remains the same as in the previous three cases – those of C2 species with 100-300 sites, those with more than 300 sites and those of C3 species with less than 1000 sites (Figure 5). There were almost no extinctions prior 1948 in all three species and even the numbers of their extinct populations during the communist era were very small (Figure 5).

Both species of *Platanthera* (*P. bifolia* and *P. chlorantha*) occur in meadows, shrubs as well as in bright forests. They do not seem to be directly threatened globally in the Czech Republic, but can be threatened locally by destruction of their natural habitats. They often co-occur in their localities and produce hybrids (Průša 2005). Despite the low decline in the Czech Republic, the numbers of their living populations in the UK and Estonia suffered a strong decline (Kull and Hutchings 2006). *Dactylorhiza majalis* is one of the most abundant species in the Czech Republic now (Kaplan et al. 2017), which typically occurs in wet meadows. It is a natural hybrid between other *Dactylorhiza* species (*D. fuchsii*, *D. maculata*, *D. incarnata* and maybe others) and therefore it is characterised by high variability in appearance (Průša 2005). Although the absolute number of living populations of *D. majalis* is not low, their rate of decline is worrisome (Janečková et al. 2006) and is highest in the second half of 20th century (Kaplan et al. 2017, Figure 5). The threats include mainly destruction of its natural habitats, such as drainage and overgrowing by other plants. A proper management is the most important factor for keeping its localities alive. Janečková et al. (2006) stated that the most appropriate management for *D. majalis* is mowing at least once a year after fruiting.

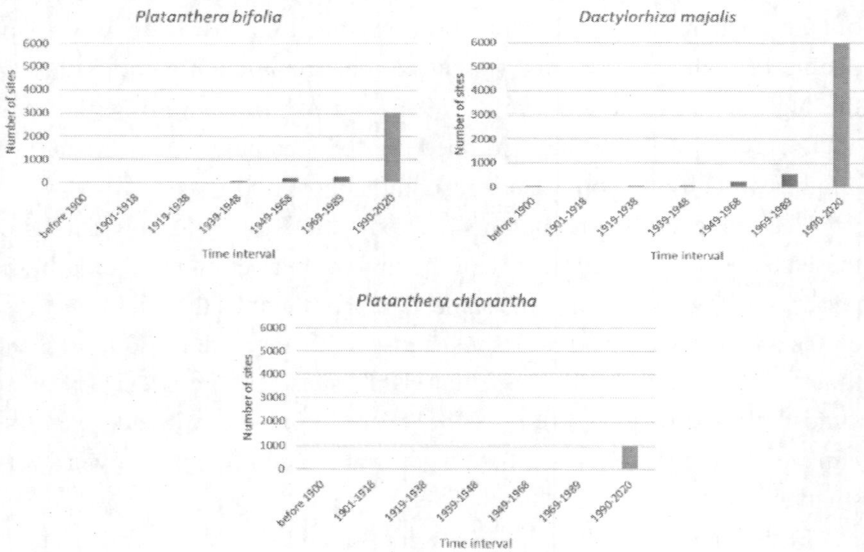

Figure 5. Numbers of sites of the three species in the C3 category (endangered species) that were found historically at 1000 and more sites for the six studied periods: blue columns are the numbers of sites that were occupied by orchids for the last time during the given interval and then not anymore, so the height of the column indicates the rate of decline of the numbers of living populations in the particular interval. The green bars indicate the numbers of living populations in the last interval, which we consider as living currently.

C4 Species

Figure 6 shows the results for the five species that are currently classified as less threatened (C4). The percentage of survival of all mentioned species is reaching 90% and the total number of living populations in the study period (1990-2020) is around 2000, the highest for *Epipactis helleborine* (4404 living populations).

The tendency for the rates of extinction of local populations of these species within sites to grow in time with the highest extinction rate in the period 1969-1989, remains the same as in the case of C2 and C3 species, which were found historically at more than 100 sites (Figure 6). There were almost no extinctions prior 1948 in all five species in this group (C4) and even the numbers of their extinct local populations within sites during the communist era were very small (Figure 6).

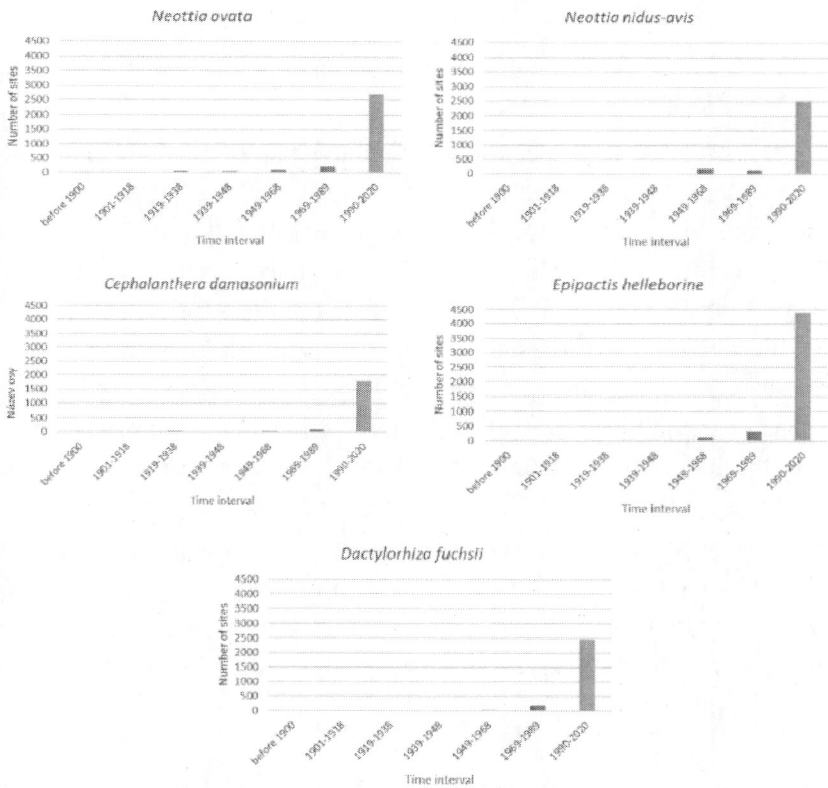

Figure 6. Numbers of sites of the five species in the C4 category (threatened species) that were found historically at their sites for the six studied periods: blue columns are the numbers of sites that were occupied by orchids for the last time during the given interval and then not anymore, so the height of the column indicates the rate of decline of the numbers of living populations in the particular interval. The green bars indicate the numbers of living populations in the last interval, which we consider as living currently.

Low decline in time of all species in this group was found also in the UK and Estonia (Kull and Hutchings 2006). It implies that these species are not directly threatened in the Czech flora (Průša 2005), but particular attention should be paid to *Cephalanthera damasonium* and *Dactylorhiza fuchsii*, as they are more prone to changes in habitats and their conditions than the other C4 species. In this group, all species can be found in open as well as in forest habitats.

CONCLUSION

Generally, for most species studied in this paper, the numbers of sites that are still occupied are much greater than those where the local populations went extinct somewhere in the period from the beginning of data included in the database until 1989. The only exceptions are the two species with very low total numbers of sites reported in the database (less than 100): *Pseudorchis albida* and *Dactylorhiza viridis*. In Štípková and Kindlmann (2021b), the opposite was true: the numbers of sites, where the local populations survived until now were comparable with those, where the local populations went extinct somewhere in the period from the beginning of data included in the database until 1990. There existed some exceptions, however: (i) *Ophrys apifera*, for which we do not have a plausible explanation; (ii) two species with relatively many historically recorded sites: *Gymnadenia densiflora*, *Dactylorhiza incarnata*; and (iii) two species with extremely many historically recorded sites: *Anacamptis morio* and *Orchis mascula*.

This all means that species with many historically recorded sites suffer a very low extinction rate of the local populations at the sites studied, which may be because these species are widely represented and do not have large demands on site quality. Species with few historically recorded sites, analysed in Štípková and Kindlmann (2021b), plus the two species with few historically recorded sites, analysed here (*Pseudorchis albida* and *Dactylorhiza viridis*), may suffer higher rates of extinction of local populations at the sites because of their large demands or some special requirements on site quality. Whether this hypothesis is true or not remains to be tested and the factors causing the extinctions of local populations at the sites determined.

Another conspicuous difference between the species analysed here and those analysed in Štípková and Kindlmann (2021b)is that in the former, the extinction rates of local populations at the sites were largest during the last two periods considered, i.e., during the communist era in 1948-1989. On the contrary, in species analysed by Štípková and Kindlmann (2021b), no such tendency was observed. It seems that this was caused by some events that occurred during the communist era, which affected even the sites of less ecologically demanding species.

To summarize: it appears that possibly the species with low ecological demands on their sites may have a lot of historically recorded sites and a low extinction rate of local populations at the sites, while species requesting special site conditions may occur only in few historically recorded sites and even in these they may suffer large extinction rates of local populations.

Almost all species declineswere caused by destruction of their natural habitats. From the conservation point of view it may mean that extinction-prone species (those with a low number of sites) should be paid a special attention in terms of maintenance of good conditions of their sites, while site quality of more abundant species is not so important, as extinction is less likely. However, the fact that in these species extinctions of their local populations occurred mainly during the communist era may indicate that during that period there were large disturbances, like conversion into building sites, especially large inputs of pesticides or artificial fertilizers into the adjacent fields, or similar, which caused extinction even of these sites.

About one third of species studied here (11species) occur in open habitats, such as grasslands, meadows and/or pastures. These habitats seem to have been affected to a greater extent by changes in agricultural and management practices in the past than species typically occurring in forests (29 species in this study). However, some species can be found both in open and forest habitats. Forests were much less affected by the changes in the agricultural practices and by deforestation (Skaloš et al.

2012) and the orchid flora in these areas was not affected to the same degree as that in grassland areas.

Acknowledgments

We thank Tony Dixon for careful editing the language and improving the quality of presentation.

References

Adams, Arthur E., and Jan S. Adams. Men Versus Systems. *Agriculture in the USSR, Poland, and Czechoslovakia*. New York: Free Press, 1971.

Bilz, Melanie, Shelagh P. Kell, Nigel Maxted, and Richard V. Lansdown. *European Red List of Vascular Plants*. Luxembourg: Publication Office of the European Union, 2011.

Cribb, Phill J., Shelagh P. Kell, Kingley W. Dixon, and Russell L. Barrett. "Orchid conservation: A global perspective." In *Orchid Conservation,* edited by Dixon, Kingley W., Shelagh P. Kell, Rusell L. Barrett, and Phill J. Cribb, 1–2. Kota Kinabalu: Natural History Publications, 2003.

Danihelka, Jiří, Jindřich Chrtek, and Zdeněk Kaplan. "*Checklist of vascular plants of the Czech Republic*." *Preslia* 84 (2012): 647–811.

Darwin, Charles. The Various Contrivances by which Orchids are Fertilized by Orchids. London: John Murray: London, 1862.

Djordjević, Vladan, Dmitar Lakušić, Slobodan Jovanović, and Vladimir Stevanović. "Distribution and Conservation Status of Some Rare and Threatened Orchid Taxa in the Central Balkans and the Southern Part of the Pannonian Plain." *Wulfenia* 24 (2017): 143–62.

ESRI. *ArcGIS Desktop: Release 10. Redlands: Environmental Systems Research Institute*, 2012.

Fick, Stephen E., and Robert J. Hijmans. "WorldClim 2: new 1km spatial resolution climate surface for global land areas." *International Journal of Climatology* 37, no. 12 (2017): 4302–15.

Fischer, Markus, and Jürg Stöcklin. "Local extinction of plants in remnants of extensively used calcareous grasslands 1950-1985." *Conservation Biology* 11 (1997): 727–37.

Gaston, Kevin J., and Tim M. Blackburn. *Patterns and Process in Macroecology*. Cambridge: Blackwell Science Ltd, 2000.

Govaerts, Rafaël. "World Checklist of Orchidaceae." Accessed January 31, 2020. http://wcsp.science.kew.org/.

Grulich, Vít. "Red List of Vascular Plants of the Czech Republic: 3rd edition." *Preslia* 84 (2012): 631–45.

Grulich, Vít, and Karel Chobot. "The Red List of Vascular Plants of the Czech Republic." *Příroda* 35 (2017): 75–132.

Henle, Klaus, Didier Alard, Jeremy Clitherow, Paul Cobb, Les Firbank, Tiiu Kull, Davy McCracken, Robin F. A. Moritz, Jari Niemela, Michael Rebane, Dirk Wascher, Allan Watt, and Juliette Young. "Identifying and Managing the Conflicts Between Agriculture and Biodiversity Conservation in Europe – A review." *Agriculture Ecosystems and Environment* 124 (2008): 60–71.

Hoffmann, M., T. M. Brooks, G. A. B. da Fonseca, C. Gascon, A. F. A. Hawkins, R. E. James, P. Langhammer, R. A. Mittermeier, J. D. Pilgrim, A. S. L. Rodrigues, and J. M. C. Silva. "Conservation Planning and the IUCN Red List." *Endangered Species Research* 6 (2008): 113–25. https://doi.org/10.3354/esr00087.

IPBES. *Global Assessment Report on Biodiversity and Ecosystem Services*. Accessed April 16, 2021. https://www.ipbes.net/global-assessment.

Jacquemyn, Hans, Rein Brys, and Michael J. Hutchings. "Biological Flora of the British Isles: *Epipactis palustris*." *Journal of Ecology* 102, no. 5 (2014): 1341–55. https://doi.org/10.1111/1365-2745.12287.

Janečková, Petra, Kateřina Wotavová, Iva Schödelbauerová, Jana Jersáková, and Pavel Kindlmann. "Relative Effects of Management and Environmental Conditions on Performance and Survival of Populations of a Terrestrial Orchid, *Dactylorhiza majalis*." *Biological Conservation* 129, no. 1 (2006): 40–49.

Jersáková, Jana, Tamara Malinová, Kateřina Jeřábková, and Stefan Dötterl. "Biological Flora of the British Isles: *Pseudorchis albida* (L.) Á. & D. Löve." *Journal of Ecology* 99, no. 5 (2011): 1282–98. https://doi.org/10.1111/j.1365-2745.2011.01868.x.

Kaplan, Zdeněk, Jiří Danihelka, Kateřina Šumberová, Jindřich Chrtek, O. Rotreklová, Libor Ekrt, J. Štěpánková, Vojtěch Taraška, Bohumil Trávníček, Jan Prančl, M. Ducháček, Michal Hroneš, Lucie Kobrlová, David Horák, and Jan Wild. "Distributions of Vascular Plants in the Czech Republic. Part 5." *Preslia* 89 (2017): 333–439.

Kull, Tiiu, and Michael J. Hutchings. "A Comparative Analysis of Decline in the Distribution Ranges of Orchid Species in Estonia and the United Kingdom." *Biological Conservation* 129 (2006): 31–39.

Kull, Tiiu, Pavel Kindlmann, Michael J. Hutchings, and Richard Primack. "Conservation biology of orchids: Introduction to the special issue." *Biological Conservation* 129 (2006): 1–3.

Kull, Tiiu, Toomas Kukk, Malle Leht, Heljo Krall, Ülle Kukk, Kalevi Kull, and Vilma Kuusk. "Distribution Trends of Rare Vascular Plant Species in Estonia." *Biodiversity and Conservation* 11 (2002): 171–96.

Kull, Tiiu, Ulvi Selgis, Miguel V. Pecina, Mirjam Metsare, Aigi Ilves, Kadri Tali, Kale Sepp, Kalevi Kull, and Richard P. Shefferson. "Factors Influencing IUCN Threat Levels to Orchids Across Europe on the Basis of National Red Lists." *Ecology and Evolution* 2016, no. 6 (2016): 6245–65.

Ljubka, T., Á. Lovas-Kiss, A. Takács, and A. V. Molnár. "*Epipactis albensis* (Orchidaceae) in Ukraine – New Data on Occurrence and Ecology." *Acta Botanica Hungarica* 56, no. 3-4 (2014): 399–408. https://doi.org/10.1556/abot.56.2014.3-4.14.

Maes, Dirk, Nick J. B. Isaac, Colin A. Harrower, Ben Collen, Arco J. van Strien, and David B. Roy. "The Use of Opportunistic Data for IUCN Red List Assessments: IUCN Red Listing Using Opportunistic Data." *Biological Journal of the Linnean Society* 115, no. 3 (2015): 690–706.

Meekers, Tine, Michael J. Hutchings, Olivier Honnay, and Hans Jacquemyn. "Biological Flora of the British Isles: *Gymnadenia conopsea* s.l." *The Journal of Ecology* 100, no. 5 (2012): 1269–88. https://doi.org/10.1111/j.1365-2745.2012.02006.x.

Molnár, Attila V. *Atlas der Orchideen Ungarns*. Budapest: Kossuth Kaidó, 2011. [*Atlas of the orchids of Hungary*]

Molnár, Attila V., and Gábor Sramkó. "*Epipactis albensis* (Orchidaceae): A New Species in the Flora of Romania." *Biologia* 67 (2012): 883–88.

Neiland, Mary Ruth M., and Christopher C. Wilcock. "Fruit Set, Nectar Reward, and Rarity in the Orchidaceae." *American Journal of Botany* 85 (1998): 1657–71.

Palacký University Olomouc. *Climatic Conditions of the Czech Republic*. Accessed January 6, 2020. https://geography.upol.cz/soubory/lide/smolova/GCZ/GCZ_Klima.pdf.

Průša, David. *Orchids of the Czech Republic* [*Orchideje Českére publiky*]. Brno: Computer Press, 2005.

Sharrock, Suzanne, and Meirion Jones. *Conserving Europe's Threatened Plants: Progress towards Target 8 of the Global Strategy for Plant Conservation*. Richmond: Botanic Gardens Conservation International, 2009.

Skaloš, Jan, Barbora Engstová, Ivana Trpáková, Markéta Šantrůčková, and Vilém Podrázský. "Long-Term Changes in Forest Cover 1780–2007 in Central Bohemia, Czech Republic." *European Journal of Forest Research* 131, no. 3 (2012): 871–84. https://doi.org/10.1007/s10342-011-0560-y.

Stewart, Joyce. *The Conservation of European Orchids. Nature and Environment*, No. 57. Strasbourg: Council of Europe Press, 1992.

Štípková, Zuzana, and Pavel Kindlmann. "Factors Determining the Distribution of Orchids in the Wild – A Review with Examples from the Czech Republic." *European Journal of Environmental Sciences* (2021a), accepted.

Štípková, Zuzana, and Pavel Kindlmann. "Orchid Extinction over the Last 150 Years in the Czech Republic." *Diversity* 13 (2021b): 78.

Štípková, Zuzana, Spyros Tsiftsis, and Pavel Kindlmann. "Pollination Mechanisms are Driving Orchid Distribution in Space." *Scientific Reports* 10 (2020): 850. https://doi.org/10.1038/s41598-020-57871-5.

Štípková, Zuzana, Spyros Tsiftsis, and Pavel Kindlmann. "How Did the Agricultural Policy During the Communist Period Affect the Decline in Orchid Biodiversity in Central and Eastern Europe?" *Global Ecology and Conservation* 26 (2021): e01498.

Swarts, Nigel D., and Kingsley W. Dixon. "Terrestrial Orchid Conservation in the Age of Extinction." *Annals of Botany* 104 (2009): 543–56. https://doi.org/10.1093/aob/mcp025.

Tsiftsis, Spyros, Ioannis Tsiripidis, and Panayiotis Trigas. "Identifying Important Areas for Orchid Conservation in Crete." *European Journal for Environmental Sciences* 1, no. 2 (2011): 28–37.

Walker, Kevin J. *Epipactis atrorubens* (Hoffm.) Besser. *Dark-red Helleborine. Species Account.* Botanical Society of Britain and Ireland, 2015.

Wädekin, Karl-Eugen. *Agrarian Policies in Communist Europe. A Critical Introduction*. Totowa: Allanheld, Osmun Publishers, 1982.

Wotavová, Kateřina, Zuzana Balounová, and Pavel Kindlmann. "Factors Affecting Persistence of Terrestrial Orchids in Wet Meadows and Implications for Their Conservation in a Changing Agricultural Landscape." *Biological Conservation* 118 (2004): 271–79.

Zamin, Tara J., Jonathan E. M. Baillie, Rebecca M. Miller, Jon P. Rodriguez, Ana Ardid, and Ben Collen. "National Red Listing Beyond the 2010 Target: Addressing National Red-List Gaps." *Conservation Biology* 24, no. 4 (2010): 1012–20.

Reviewed by:

Spyros Tsiftsis

Department of Forest and Natural Environment Sciences, International Hellenic University, Drama, Greece.

In: Orchidaceae
Editor: Vladan Djordjević

ISBN: 978-1-68507-189-9

Chapter 3

ORCHIDS OF SERBIA: TAXONOMY, LIFE FORMS, POLLINATION SYSTEMS, AND PHYTOGEOGRAPHICAL ANALYSIS

Vladan Djordjević[1,*], Marjan Niketić[2] and Vladimir Stevanović[1,3]

[1]University of Belgrade, Faculty of Biology, Institute of Botany and Botanical Garden, Belgrade, Serbia

[2]Natural History Museum, Belgrade, Serbia

[3]Serbian Academy of Sciences and Arts, Belgrade, Serbia

ABSTRACT

The Balkan Peninsula is one of the most important centers of orchid diversity in Europe. However, the orchid flora in the central Balkans and southern part of the Pannonian Plain is still not sufficiently known. Therefore, this chapter reviews the orchid flora of Serbia, considering its taxonomic structure, life forms, and pollination systems. Moreover, detailed insight into the chorological groups of individual taxa is provided. Long-term personal field investigations, checking and

[*] Corresponding Author's E-mail: vdjordjevic@bio.bg.ac.rs.

revision of herbarium material, and relevant published sources were used to complete the overview of orchid taxa present in Serbia. In total, 72 orchid species and subspecies belonging to 22 genera have been recorded in Serbia, whereas the presence of three taxa is doubtful. Taxonomic analysis shows that the most species-rich genera are *Epipactis* and *Dactylorhiza* (10 taxa each), followed by *Anacamptis* (nine taxa), then *Orchis* (eight taxa) and *Ophrys* (seven taxa). Examination of the life form spectrum indicates that orchids with ovoid tubers are the most numerous, followed by rhizomatous orchids, whereas there are slightly fewer taxa belonging to the group of orchids with palmately lobed and fusiform tubers. As far as pollination systems are concerned, the deceptive system is present in the greatest number of orchids, the rewarding system is the second largest system, whereas self-pollinating orchids are less numerous. Phytogeographical analysis shows domination of Central European and Mediterranean-Submediterranean taxa, followed by taxa from the Eurasian and boreal chorological groups. The importance of the Balkan endemics *Himantoglossum calcaratum* subsp. *calcaratum* and *Dactylorhiza cordigera* subsp. *bosniaca* and that of three subendemics of the Carpathians and the Balkans (*Gymnadenia frivaldii*, *Dactylorhiza maculata* subsp. *transsilvanica,* and *Dactylorhiza cordigera* subsp. *cordigera*) are especially emphasized. In addition, this synthesis underlines the fact that the study area provides suitable conditions for the survival of some orchid species that have southern or northern limits of their distribution in this part of southeastern Europe. Bearing in mind the great diversity of relief, vegetation, climatic conditions, geological substrates, and soil characteristics, as well as the action of historical factors, we assume that the number of orchid species in Serbia could be higher and expect that new taxa will eventually be recorded. The chapter highlights the importance of the role played by orchid life history strategies in studies of diversity patterns, as well as the need for detailed future research on the taxonomy, chorology, ecology, and conservation of orchids in Serbia.

Keywords: Orchidaceae, Balkan Peninsula, orchid flora, taxonomic analysis, life history strategies, chorology, biodiversity

Introduction

The family Orchidaceae is one of the largest and most diverse of plant families, with an estimated number of approximately 28,000 species within 749 genera (Christenhusz and Byng 2016; Chase et al. 2017). Orchids comprise approximately 8% of all vascular plants, are native to all continents except Antarctica, and inhabit almost all terrestrial ecosystems except the driest deserts, salt marshes and agricultural lands (Hágsater and Dumont 1996; Givnish et al. 2016). Known for their attractive flowers, complex biotic interactions, and intriguing ecology, they have long fascinated scientists and nature lovers. The family Orchidaceae is divided into five subfamilies: Apostasioideae, Vanilloideae, Cypripedioideae, Orchidoideae, and Epidendroideae (Chase et al. 2015). Separation of the first subfamilies of orchids occurred in the early Cenozoic, 56 to 52 million years ago (Ramirez et al. 2007). The diversification of the Orchidaceae, i.e., separation of the subfamilies Orchidoideae and Epidendroideae, occurred in the Eocene, 53 to 49 million years ago (Gustafsson et al. 2010).

The extraordinary diversity and species richness of orchids has been linked in many studies to their specialized interactions with mycorrhizal fungi and pollinators, evolution of pollinia, deceptive pollination, pollination by euglossine bees or Lepidoptera epiphytism and associated traits such as Crassulacean acid metabolism (CAM), and predominant distribution in the tropics (Gravendeel et al. 2004; Cozzolino and Widmer 2005; Waterman and Bidartondo 2008; Givnish et al. 2015). Orchids show a great diversity of life history traits, not only in their development and morpho-anatomical and physiological adaptations, the type of their underground organs, and the degree of their specialization with respect to pollinators and mycorrhizal symbionts, but also in the degree of their habitat specialization (Whigham and Willems 2003; Djordjević and Tsiftsis 2020; Li et al. 2021). In general, there are orchids that have underground root systems (terrestrial geophytes and non-geophytes) and orchids that grow on the branches and trunks of the host tree – epiphytic orchids (Taylor et al. 2021). Recent studies examined how environmental factors influence the distribution of orchids of specific life forms and pollination systems and emphasized importance of the role played by life history types in determining the patterns of orchid diversity (Djordjević 2018; Tsiftsis et al. 2019b; Štípková et al. 2020, 2021; Taylor et al.

2021). It is also known that a large number of orchid taxa are at risk of extinction, mostly due to loss of the natural habitats in which they grow or due to other negative human and global influences such as illegal collecting (biological resource use), fragmentation, intensive agriculture, habitat disturbance, urbanization, pollution, climate change, and modifications of natural systems (Wraith and Pickering 2018; Jakubska-Busse et al. 2021).

Although orchids have a cosmopolitan distribution, occurring from Alaska and northern Sweden in the north to Tierra del Fuego and Macquarie Island in the south almost to the extreme southern and northern limits of vegetation distribution, their greatest diversity is achieved in the tropics (Dressler 1981; Gladkova 1982; Pridgeon et al. 1999). The southern region of Central America and the northwestern region of South America have the greatest number of orchid taxa (Hágsater and Dumont 1996). Dressler (1981) stated that 8266 species from 306 genera are distributed in the tropical part of America and 6800 species from 250 genera in Southeast Asia, while 3131 species from 134 genera have been recorded in the tropical part of Africa and Madagascar. There are fewer orchid taxa in the temperate zone than in tropical regions. About 900 orchid species from 75 genera grow in the temperate zone of the Northern Hemisphere and about 500 species from 40 genera in the southern temperate zone, while about 170 orchid species have been recorded in North America (from Mexico to Canada) (Dressler 1981; Gladkova 1982).

More than 300 orchid species occur in Europe, North Africa, and the Middle East (Hágsater and Dumont 1996). According to Buttler (1991), 250 taxa from 36 genera of orchids grow in Europe. The taxonomic diversity of the family Orchidaceae gradually increases from the north to the south of Europe, being greatest in the Mediterranean areas (Djordjević and Tsiftsis 2020). The Balkan Peninsula, the Alps, Sicily, and the Caucasus stand out as important centers of orchid diversity in Europe (Hágsater and Dumont 1996). However, the most important center of orchid diversity in Europe is the eastern Mediterranean, i.e., the area around the Aegean Sea (Hágsater and Dumont 1996; Tsiftsis et al.

2019b; Tsiftsis 2020). A large number of orchids in Europe have also been recorded in Italy (189 species and subspecies) (GIROS 2009); in Turkey (170 species and subspecies) (Kreutz and Çolak 2009); in France, Belgium, and Luxembourg (134 species and subspecies) (Vogt-Schilb et al. 2015); on the Iberian Peninsula (122 species and subspecies) (Bernardos et al. 2007); and in Romania (71 species and 14 subspecies) (De Angelli and Anghelescu 2020). The island of Corsica stands out: 77 orchid species were recorded there, with a density of 88 orchid species per 1000 km^2 (Vogt-Schilb et al. 2016). In contrast, lower numbers of orchid species were recorded in the United Kingdom (50 taxa) (Kull and Hutchings 2006); in the Netherlands (37 taxa) (Jacquemyn et al. 2005a); and in Estonia (36 taxa) (Kull and Hutchings 2006).

The exact number of orchid taxa on the Balkan Peninsula is not known, but it is estimated at about 220-230. Among the Balkan countries, Greece has the largest number of orchid taxa – 194 species and subspecies (Tsiftsis and Antonopoulos 2017; Tsiftsis and Tsiripidis 2020). According to Nikolić (2019), there are 113 orchid species and subspecies in Croatia, while 88 taxa have been recorded in Albania (Vangjeli et al. 2000; Barina et al. 2018). They are followed by Montenegro, with about 80 orchid taxa (Adamović 1913; Hayek 1933; Parolly 1992; Stevanović et al. 1995; Pulević 2005); Bosnia and Herzegovina, with about 75 taxa (Pantocsek 1874; Beck-Mannagetta 1886; Hayek 1933; Šoljan et al. 2014; Šabanović et al. 2020); Slovenia, with 73 taxa (Dolinar 2015); Bulgaria, with about 70 taxa (Assyov and Petrova 2012); and North Macedonia, with about 70 taxa (Bornmüller 1925; Hayek 1933; Ponert 2014).

Although the Balkan Peninsula is one of the most important centers of orchid diversity in Europe, research on orchids in the central Balkans and the southern part of the Pannonian Plain, including Serbia, has been carried out only partially and with a great discontinuity in time. The first data on orchids in Serbia were published by Ami Boué (Boué 1840) in La Turquie d' Europe, in which he noted the presence of eight orchid taxa in the flora of Serbia. Later, August Grisebach (Grisebach 1846) and Rumy (1846) also noted some orchid taxa in Serbia. In addition, reference should be made to the orchid data of Josif Pančić from the middle of the 19th century (Pančić 1856, 1865, 1874, 1884). Pančić initially noted the presence of 33 species and five varieties of orchids from 12 genera (Pančić 1856). He described *Orchis maculata* var. *ochrantha* Pančić (Pančić 1856), which is now considered to be a synonym of the officially accepted taxon *Dactylorhiza maculata* subsp. *transsilvanica* (Schur) Soó (WCSP 2020). In the work *Flora in the Environs of Belgrade: According to the Analytical Method*, Pančić (1865) subsequently recorded the occurrence of 24 orchid species from 10 genera, while in the work *Flora of the Principality of Serbia* he mentioned 36 orchid species from 16 genera (Pančić 1874). The first chorological data on orchids in the environs of Novi Sad and Fruška Gora were given by Schulzer et al. (1866) and Zorkóczy (1896). In the second half of the 19th century, data on orchids in the environs of Niš were published by Sava Petrović (Petrović 1882, 1885).

Data on the distribution of orchids in Serbia from the middle of the 19th century to the first half of the 20th century were published by the following authors: Schneller (1858), Godra (1872), Borbás (1878), Adamović (1892, 1898, 1899, 1908, 1909), Wettstein (1892), Ničić (1894), Jurišić (1901, 1913, 1923), Beck-Mannagetta (1903), Košanin (1910), Prodán (1916), Novak (1926), Bornmüller (1928), Majstorović (1929), Soó (1929), Markgraf (1932), and Rechinger (1935). Fritsch (1909) listed a significant number of orchids from Serbia, and he described *Orchis* × *serbica* H.Fleischm, on the basis of material from Vlasina collected by M. Dimitrijević in 1896. Today, this taxon is accepted under the name *Dactylorhiza* × *serbica* (H. Fleischm.) Soó (WCSP 2020) because it is a natural hybrid between *Dactylorhiza incarnata* (L.) Soó and *D. saccifera* (Brongn.) Soó. Moreover, Hayek (1933) published *Prodromus Florae Peninsulae Balcanicae*, in which he stated that 61 orchid taxa from 18 genera grow in Serbia.

Significant contributions to knowledge about the chorology and ecological preferences of orchids in Serbia can be found in floristic and vegetation studies and dissertations published from the middle to end of the 20th century (Gajić 1989; Gajić et al. 1992; Stevanović et al. 1991;

Lakušić 1993). Intensive vegetation research was carried out during this period, and a large body of data on orchids can be found in numerous phytocenological relevés. An important source of information on the distribution of orchids is material deposited in various herbariums, namely those of the University of Belgrade (BEOU); the Museum of Natural History in Belgrade (BEO); the Department of Biology and Ecology, Faculty of Sciences, University of Novi Sad (BUNS); the Institute for Nature Conservation of the Vojvodina Province (PZZP); and the Faculty of Science and Mathematics, Department of Biology and Ecology, University of Niš (HMN). Here it is especially necessary to mention Stanija Čolović, who published the first study on orchids of Mt. Fruška Gora (Čolović-Parabućski 1958). The presence of 15 orchid taxa was indicated in that study, which was the first specialized scientific work on orchids in Serbia. Nikola Diklić and Vojislav Nikolić from the Natural History Museum in Belgrade have contributed greatly to knowledge about the chorology and ecology of orchids in Serbia. They are noted for their extensive herbarium material and significant published floristic works. Diklić published the most detailed available survey of orchids (encompassing 54 species and 22 subspecies of 22 genera) as part of the eighth volume of the Flora of SR Serbia (Diklić 1976), which contained a detailed description of taxa and identification keys, as well as data on the distribution and ecological preferences of orchids.

Also worth mentioning are a work on new orchid infrataxa in the flora of Vojvodina (Obradović et al. 1983), a study of taxa from the *Anacamptis palustris* group (Boža et al. 1988), and a work on new findings of orchids in Vojvodina (Boža and Butorac 1981). Of particular importance is the paper *Orchids in the southern parts of Serbia: some recent findings* (Hill 1985), in which the author recorded the occurrence of 28 orchid taxa in southern, eastern and southeastern Serbia. Another publication that should be singled out is *Diversity of the vascular flora of Yugoslavia with an overview of species of international importance* (Stevanović et al. 1995), in which the authors gave an overview of the orchids of Serbia, the regions and subregions of Serbia in which the species are represented, and their affiliation with a particular habitat type according to the CORINE classification. The first paper on the distribution and ecological preferences of orchids in western Serbia (*Orchids of the Valjevo region*) was published in 1997 (Djordjević 1997), while Dragiša Savić wrote the first master's thesis on orchids of Mt. Fruška Gora (Savić 1998). Significant floristic works followed, providing new data on the distribution of certain orchids in Serbia (Karadžić et al. 2000; Tomović et al. 2007; Savić et al. 2008; Ponert 2014).

In recent years, several new orchid taxa and nothotaxa have been recorded for the first time in the flora of Serbia: *Epipactis purpurata* Sm. (Djordjević et al. 2010); *Neotinea* × *dietrichiana* (Bogenh.) H. Kretzschmar, Eccarius & H. Dietr. (Djordjević et al. 2012); *Dactylorhiza fuchsii* (Druce) Soó subsp. *fuchsii* (Djordjević et al. 2014); *Epipactis distans* Arv.-Touv., *E. pontica* Taubenheim and *E. leptochila* subsp. *neglecta* Kümpel (Djordjević et al. 2016a); *E. muelleri* Godfery (Djordjević 2016); *Orchis* × *angusticruris* Franch. (Radak et al. 2016, 2019a); *Orchis* × *beyrichii* (Rchb.f.) A.Kern. and *Anacamptis* × *timbali* (Velen.) H. Kretzschmar, Eccarius & H.Dietr. (Radak 2019a, 2019b); *Epipactis tallosii* A.Molnár & Robatsch (Süveges et al. 2019); and *Orchis provincialis* Balb. ex Lam. & DC. (Djordjević et al. 2021). In addition, data on the distribution and conservation status of 10 orchid species in Serbia have also been published (Djordjević et al. 2017).

Among recent ecological studies of orchids in Serbia, a study on ecological niches of orchids in the Valjevo Mountains of northwestern Serbia stands out for finding that habitats on serpentine substrates play an important role in the survival of numerous orchid species previously listed as characteristic of carbonate habitats (Djordjević 2012; Djordjević et al. 2016c). In addition, studies have been published on the distribution, abundance, and composition of orchids of the herbaceous vegetation type (Djordjević et al. 2016b) and on forest orchids (Djordjević et al. 2020a) in western Serbia. Orchid species richness and composition in relation to geological substrates (Djordjević and Tsiftsis 2019) and vegetation types (Djordjević et al. 2020b) have also been studied.

An especially noteworthy recent publication is An Annotated Checklist of Vascular Flora of Serbia 1 (Niketić and Tomović 2018), the authors of which indicate the presence or absence of orchids in basic territorial units of Serbia (Central Serbia, Vojvodina, Kosovo and Metohija) (Djordjević et al. 2018). A number of floristic and chorological studies of orchid taxa in Serbia have been published since then (Niketić et al. 2018, 2020; Radak et al. 2019a, 2019b; Tomović et al. 2020). These studies contributed significantly to a better awareness of the occurrence and distribution of the orchid flora in Serbia. In addition, Niketić et al. (2018) proposed three new nomenclatural combinations for representatives of the family Orchidaceae: *Dactylorhiza saccifera* subsp. *macrostachys* (Tineo) Niketić & Djordjević, comb. nov., *Himantoglossum calcaratum* subsp. *rumelicum* (H. Baumann & R. Lorenz) Niketić & Djordjević, comb. nov., and *Ophrys sphegodes* subsp. *taurica* (Aggeenko) Soó ex Niketić & Djordjević, comb. nov. Three PhD theses on orchids in Serbia have been published in recent years: Spatial distribution and ecology of orchids (Orchidaceae) of western Serbia (Djordjević 2018), Morphological variability of species of the genus *Anacamptis* Rich. (Orchidoideae, Orchidaceae) on the Balkan Peninsula and the Pannonian Plain (Radak 2019), and Biodiversity and breeding of terrestrial orchids from Fruška Gora mountain (Dulić 2019). Their publication indicates a significant increase in the interest of Serbian scientists in study of this plant family.

Because research on the distribution and ecology of orchids in Serbia was mostly carried out within the framework of floristic or phytocoenological studies of specific areas, there is a need for a comprehensive analysis of the orchid flora in Serbia, both taxonomically and from the point of view of life history and chorology. Accordingly, the main objectives of this chapter are to provide: (i) a detailed overview of orchid taxa in Serbia; (ii) a taxonomic analysis; (iii) an analysis of orchid life forms; (iv) an analysis of pollination systems; and (v) a phytogeographical analysis of the orchid flora in Serbia.

Systematic Overview of the Orchid Flora of Serbia

The overall survey of orchid taxa occurring in Serbia given here is based on long-term personal field investigations, herbarium material, and relevant published sources. Material in the herbarium of the University of Belgrade (BEOU) and in that of the Museum of Natural History in Belgrade (BEO) was reviewed and revised. With few exceptions, the nomenclature and taxonomy in this chapter follow An Annotated Checklist of Vascular Flora of Serbia 1 (Djordjević et al. 2018).

Family Orchidaceae

Subfamily Cypripedioideae

Tribe Cypripedieae

Subtribe: Cypripediinae

Genus *Cypripedium* L.
Cypripedium calceolus L.

Subfamily Epidendroideae

Tribe Neottieae

Genus *Cephalanthera* Rich.
Cephalanthera damasonium (Mill.) Druce
Cephalanthera longifolia (L.) Fritsch
Cephalanthera rubra (L.) Rich.

Genus *Epipactis* Zinn
Epipactis atrorubens (Hoffm.) Besser
Epipactis distans Arv.-Touv.
Epipactis helleborine (L.) Crantz subsp. *helleborine*
Epipactis leptochila subsp. *neglecta* Kümpel
Epipactis microphylla (Ehrh.) Sw.
Epipactis muelleri Godfery subsp. *muelleri*
Epipactis palustris (L.) Crantz
Epipactis pontica Taubenheim
Epipactis purpurata Sm.
Epipactis tallosii A.Molnár & Robatsch subsp. *tallosii*

Genus *Limodorum* Boehm.
Limodorum abortivum (L.) Sw.

Genus *Neottia* Guett.
Neottia cordata (L.) Rich.
Neottia nidus-avis (L.) Rich.
Neottia ovata (L.) Bluff & Fingerh.

Tribe Nervilieae

Subtribe Epipogiinae

Genus *Epipogium* Borkh.
Epipogium aphyllum Sw.

Tribe Epidendreae

Subtribe: Calypsoinae

Genus *Corallorhiza* Gagnebin
Corallorhiza trifida Châtel.

Subfamily Orchidoideae

Tribe Cranichideae

Subtribe Goodyerinae

Genus *Goodyera* R.Br. in W.T.Aiton
Goodyera repens (L.) R.Br.

Subtribe Spiranthinae
Genus *Spiranthes* Rich.
Spiranthes aestivalis (Poir.) Rich.
Spiranthes spiralis (L.) Chevall.

Tribe Orchideae

Subtribe Orchidinae

Genus *Anacamptis* Rich.
Anacamptis coriophora (L.) R.M.Bateman, Pridgeon & M.W.Chase subsp. *coriophora*
Anacamptis coriophora subsp. *fragrans* (Pollini) R.M.Bateman, Pridgeon & M.W.Chase
Anacamptis laxiflora (Lam.) R.M.Bateman, Pridgeon & M.W.Chase
Anacamptis morio (L.) R.M.Bateman, Pridgeon & M.W.Chase subsp. *morio*
Anacamptis morio subsp. *caucasica* (K.Koch) H.Kretzschmar, Eccarius & H.Dietr.
Anacamptis palustris (Jacq.) R.M.Bateman, Pridgeon & M.W.Chase subsp. *palustris*
Anacamptis palustris subsp. *elegans* (Heuff.) R.M.Bateman, Pridgeon & M.W.Chase

Anacamptis papilionacea (L.) R.M.Bateman, Pridgeon & M.W.Chase subsp. *papilionacea*

Anacamptis pyramidalis (L.) Rich.

Genus *Coeloglossum* Hartm.

Coeloglossum viride (L.) Hartm.

Genus *Dactylorhiza* Neck. ex Nevski

Dactylorhiza cordigera (Fr.) Soó subsp. *cordigera*

Dactylorhiza cordigera subsp. *bosniaca* (Beck) Soó

Dactylorhiza fuchsii (Druce) Soó subsp. *fuchsii*

Dactylorhiza incarnata (L.) Soó subsp. *incarnata*

Dactylorhiza maculata (L.) Soó subsp. *maculata*

Dactylorhiza maculata subsp. *transsilvanica* (Schur) Soó

Dactylorhiza majalis (Rchb.) P.F.Hunt & Summerh. subsp. *majalis*

Dactylorhiza romana (Sebast.) Soó subsp. *romana*

Dactylorhiza saccifera (Brongn.) Soó subsp. *saccifera*

Dactylorhiza sambucina (L.) Soó

Genus *Gymnadenia* R.Br.

Gymnadenia conopsea (L.) R.Br.

Gymnadenia frivaldii Hampe ex Griseb.

Gymnadenia odoratissima (L.) Rich.

Genus *Herminium* L.

Herminium monorchis (L.) R.Br.

Genus *Himantoglossum* Spreng.

Himantoglossum calcaratum (Beck) Schltr. subsp. *calcaratum*

Himantoglossum calcaratum subsp. *rumelicum* (H. Baumann & R. Lorenz) Niketić & Djordjević

Genus *Neotinea* Rchb.f.

Neotinea lactea (Poir.) R.M.Bateman, Pridgeon & M.W.Chase

Neotinea tridentata (Scop.) R.M.Bateman, Pridgeon & M.W.Chase subsp. *tridentata*

Neotinea ustulata (L.) R.M.Bateman, Pridgeon & M.W.Chase

Genus *Nigritella* Rich.

Nigritella rhellicani Teppner & E.Klein

Genus *Ophrys* L.

Ophrys apifera Huds.

Ophrys fuciflora (F.W.Schmidt) Moench subsp. *fuciflora*

Ophrys insectifera L. subsp. *insectifera*

Ophrys scolopax subsp. *cornuta* (Steven) E.G.Camus in E.G.Camus

Ophrys sicula Tineo

Ophrys sphegodes Mill. subsp. *sphegodes*

Ophrys sphegodes subsp. *taurica* (Aggeenko) Soó ex Niketić & Djordjević

Genus *Orchis* Tourn. ex L.

Orchis mascula (L.) L. subsp. *mascula*

Orchis mascula subsp. *speciosa* (Mutel) Hegi

Orchis militaris L. subsp. *militaris*

Orchis pallens L.

Orchis provincialis Balb. ex Lam. & DC.

Orchis purpurea Huds. subsp. *purpurea*

Orchis simia Lam. subsp. *simia*

Orchis spitzelii Saut. ex W.D.J.Koch subsp. *spitzelii*

Genus *Platanthera* Rich.

Platanthera bifolia (L.) Rich.

Platanthera chlorantha (Custer) Rchb. subsp. *chlorantha*

Genus *Pseudorchis* Ség

Pseudorchis albida (L.) Á.Löve & D.Löve subsp. *albida*

Genus *Traunsteinera* Rchb.

Traunsteinera globosa (L.) Rchb.

Hybrids:

Anacamptis × *timbali* (Velen.) H.Kretzschmar, Eccarius & H.Dietr. nothosubsp. *timbali*

[= *A. coriophora* × *A. palustris* subsp. *palustris*]

Cephalanthera × *schulzei* E.G.Camus in E.G.Camus

[= *C. damasonium* × *C. longifolia*]

Dactylorhiza × *serbica* (H.Fleischm.) Soó

[= *D. incarnata* × *D. saccifera*]

Neotinea × *dietrichiana* (Bogenh.) H.Kretzschmar, Eccarius & H.Dietr. nothosubsp. *dietrichiana* [= *N. tridentata* subsp. *tridentata* × *N. ustulata* var. *ustulata*]

Ophrys × *hybrida* Pokorny ex Rchb.f.

[= *O. insectifera* subsp. *insectifera* × *O. sphegodes* subsp. *sphegodes*]

Orchis × *angusticruris* Franch. nothosubsp. *angusticruris*

[= *O. purpurea* subsp. *purpurea* × *O. simia*]

Orchis × *beyrichii* (Rchb.f.) A.Kern. nothosubsp. *beyrichii*

[=*O. militaris* subsp. *militaris* × *O. simia*]

Orchis × *hybrida* (Lindl.) Boenn. ex Rchb.

[=*O. militaris* × *O. purpurea*]

RICHNESS OF ORCHID TAXA AND TAXONOMIC ANALYSIS

The floristic composition of the orchid flora in Serbia includes three subfamilies, six tribes, 22 genera, 64 species, eight subspecies, and eight hybrid taxa (Figures 1–3). However, for five taxa (*Anacamptis coriophora* subsp. *fragrans*, *Dactylorhiza romana*, *Neotinea lactea*, *Ophrys sicula*, and *Orchis mascula* subsp. *mascula*) there are only published data, but no herbarium material or field observations confirming their occurrence in Serbia (Diklić 1976; Djordjević et al. 2018). In addition, the following three taxa are probably mislisted in the published sources: *Dactylorhiza incarnata* (L.) Soó subsp. *ochroleuca* (Wüstnei ex Boll) P.F. Hunt & Summerh., *Epipactis leptochila* (Godfery) Godfery subsp. *leptochila*, and *Orchis italica* Poir. (Djordjević et al. 2018) and are therefore not included in the analyses presented in this chapter. Furthermore, there are several erroneously reported taxa in the flora of Serbia, cases where all voucher specimens were misidentified or where reports without voucher specimens are obviously erroneous: *Anacamptis morio* subsp. *picta* (Loisel.) Jacquet & Scappat, *Himantoglossum caprinum* (M.Bieb.) Spreng, *Himantoglossum hircinum* (L.) Spreng, *Himantoglossum robertianum* (Loisel.) P.Delforge, *Nigritella nigra* (L.) Rchb.f., *Ophrys sphegodes* subsp. *atrata* (Rchb.f.) A.Bolòs, and *Orchis anthropophora* (L.) All. (Djordjević et al. 2018).

With a total of 72 orchid species and subspecies, Serbia can be considered an area of moderate orchid species richness in Europe. The lower number of orchid species and subspecies in Serbia compared to the more orchid-rich European countries can be explained by climatic conditions, i.e., the dominant presence of a temperate-continental climate in Serbia (Stevanović and Stevanović 1995) and the absence of Mediterranean climate and ecosystems typical of Mediterranean areas, where the largest number of orchids from the genera *Ophrys, Orchis*, and *Serapias* occur (Delforge 2006; GIROS 2009; Tsiftsis and Antonopoulos 2017; Tsiftsis and Djordjević 2018, 2020). The lower number of orchid species in Serbia can also be explained by the lower representation of high-altitude areas (areas higher than 1800 m a.s.l.), as well as by the fact that areas of lower altitude are under strong zoo-anthropogenic influence. In addition, the high prevalence of non-carbonate (ultramafic and silicate) geological substrates in Serbia may also be a reason for the smaller number of members of the family Orchidaceae, given the fact that in Europe areas dominated by carbonates have the greatest richness of orchids (Djordjević and Tsiftsis 2019).

The subfamily Cypripedieae is represented by one species (1.39%), Epidendroideae includes 19 species and subspecies (26.39%), while the subfamily Orchidoideae is represented by 52 species and subspecies (72.22%). The richest tribe is Orchideae, with 49 species and subspecies (68.06%), followed by Neottieae with 18 and Cranichideae with three species, while the tribes Cypripedieae, Nervilieae, and Epidendreae are represented by one species each (Figure 2).

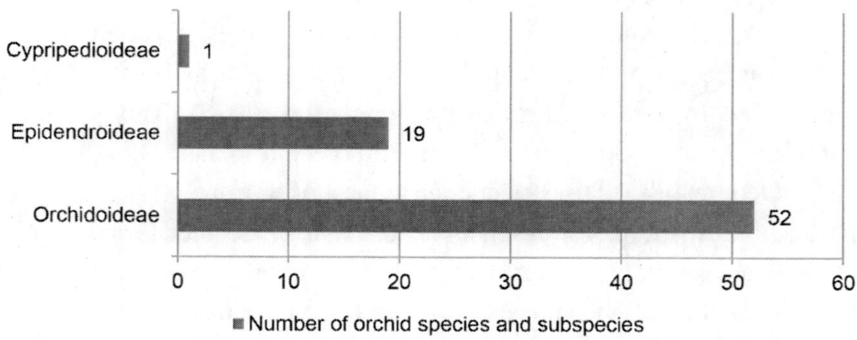

Figure 1. Representation of subfamilies in the orchid flora of Serbia.

The most taxon-rich genera are *Epipactis* and *Dactylorhiza* (with 10 taxa each), *Anacamptis* (nine taxa), *Orchis* (eight taxa), and *Ophrys* (seven taxa). The genera *Cephalanthera*, *Neottia*, *Gymnadenia*, and *Neotinea* are represented by three taxa each, while the genera *Spiranthes*, *Himantoglossum*, and *Platanthera* are represented by two taxa each. Ten genera (*Cypripedium*, *Limodorum*, *Epipogium*, *Corallorhiza*, *Goodyera*, *Coeloglossum*, *Herminium*, *Nigritella*, *Pseudorchis*, and *Traunsteinera*) are represented by a single taxon (Figure 3).

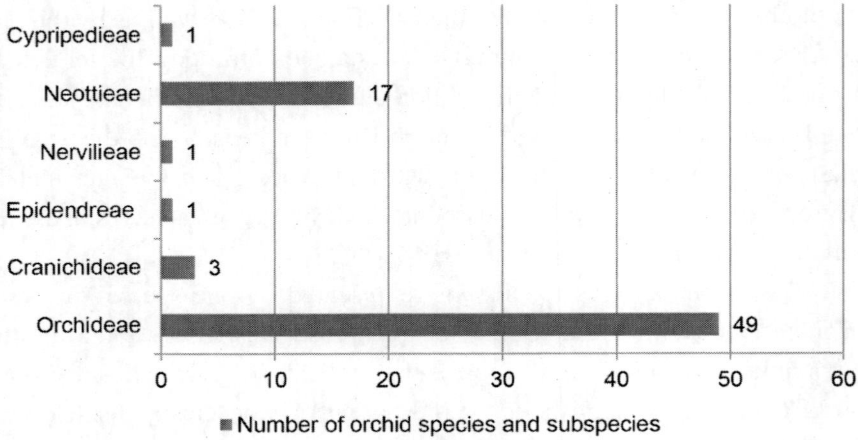

Figure 2. Representation of tribes in the orchid flora of Serbia.

The orchids of Serbia are classified into three subfamilies (Figure 1) out of a total of five subfamilies of the family Orchidaceae (Chase et al. 2015). The subfamily Orchidoideae, with four tribes (Codonorchideae, Chranichideae, Diurideae, and Orchideae) and 21 subtribes, is mainly represented by terrestrial orchids (Chase et al. 2015). This subfamily includes about 3630 species from 200 genera (Chase et al. 2017), distributed mainly in the temperate zones of Europe, Asia, America, Africa, and Australia. Representatives of the tribe Orchideae and subtribe Orchidinae dominate in Eurasia and North America, representatives of the tribe Orchideae and subtribes Disinae and Satyriinae are distributed in South Africa, while taxa from the tribe Cranichideae and subtribes Diurideae occur in southern South America (Chase et al. 2017). Species of the given subfamily are distinguished by the presence of root-stem tuberoids in most members, and by a usually erect anther, projecting above the stigma (Dressler 1981; Chase et al. 2017). Only a small number of representatives of the subfamily are epiphytic (Chase et al. 2017). In general, the results of taxonomic analysis of the orchid flora of Serbia showed that representatives of the tribe Orchideae and subtribe Orchidinae dominate in regard to the number of genera (11 out of 22) and the number of species and subspecies (68.06% of the total).

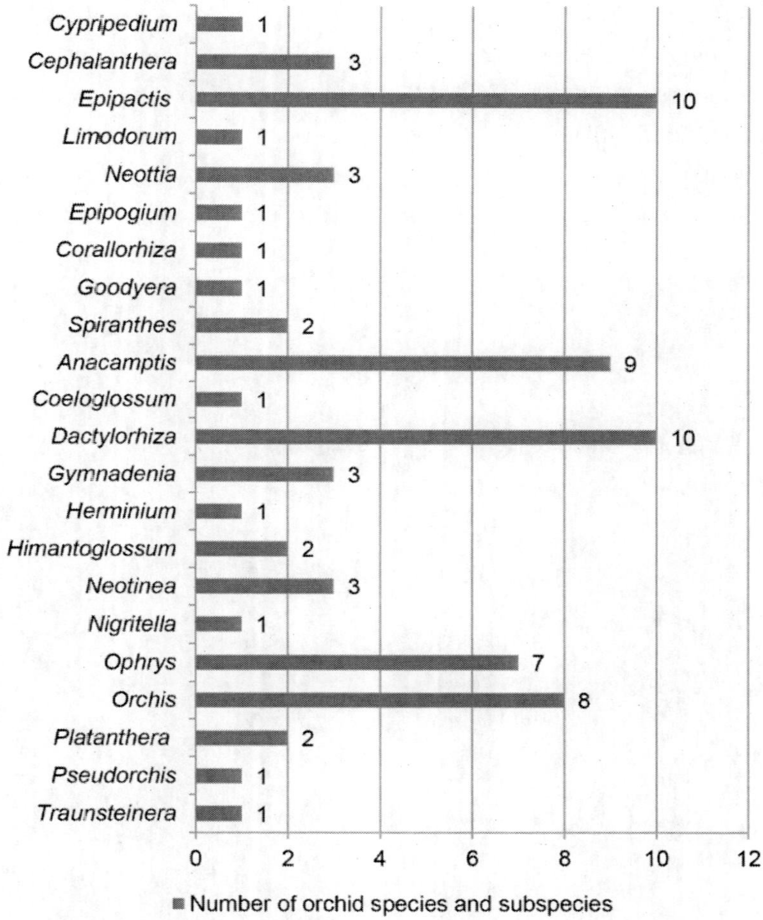

Figure 3. Representation of genera in the orchid flora of Serbia.

The tribe Orchideae includes terrestrial orchids from 62 genera (Inda et al. 2012), which are mainly distributed in the temperate zone of the Northern Hemisphere and are less abundant in tropical areas. A total of four subtribes (Brownleeinae, Coryciinae, Disinae, and Orchidinae) are included in this tribe (Chase et al. 2015). Based on phylogenetic analysis using ITS (internal transcribed spacer) sequences of nuclear ribosomal DNA, the subtribe Orchidinae is divided into 12 clades (*Anacamptis, Serapias, Ophrys, Steveniella-Himantoglossum, Neotinea, Traunsteinera-Chamorchis, Orchis, Pseudorchis-Amerorchis-Galearis-Neolindleya-*

Platanthera, Dactylorhiza, Gymnadenia, Ponerorchis-Hemipilia-Amitostigma-Neottianthe, and *Brachycorythis*) (Bateman et al. 2003). This subtribe includes about 1800 species from 47 to 60 genera (Jin et al. 2017). Orchids from nine clades have been found in Serbia, but no representatives of the clades *Serapias*, *Ponerorchis-Hemipilia-Amitostigma-Neottianthe*, or *Brachycorythis* have been recorded there.

Figure 4. Some representatives of the genus *Dactylorhiza* in Serbia: a) *Dactylorhiza cordigera* subsp. *cordigera*, b) *Dactylorhiza incarnata* subsp. *incarnata*, c) *Dactylorhiza sambucina*, d) *Dactylorhiza maculata* subsp. *maculata*, e) *Dactylorhiza maculata* subsp. *transsilvanica*, f) *Dactylorhiza saccifera* subsp. *saccifera* (photos taken from 2006 to 2020 by V. Djordjević).

The genus *Dactylorhiza* ranks first in regard to the number of orchid taxa within both the subfamily Orchidoideae and the total orchid flora of Serbia (Figure 3). The occurrence of a large number of taxa of this genus in Serbia is understandable, in view of the fact that the climate of Serbia is humid, especially in the western and southwestern parts, and due to the significant presence of silicate geological substrates, which are known for their high water-storage capacity (Djordjević and Tsiftsis 2019), as well as wet habitats (wet meadows, fens, bogs, and marshes), which are typical habitats for representatives of this genus (Averyanov 1990; Delforge 2006; Pillon et al. 2006).

Dactylorhiza species are distributed mainly in temperate and boreal regions, for the most part in Europe, but also in western and northern Asia, the Himalayas, North Africa, and Alaska (Averyanov 1990; Delforge 2006; Ståhlberg and Hedrén 2008; Vakhrameeva et al. 2008). The centers of diversity of this genus are Western Europe (Great Britain, Germany, and southern Scandinavia), the Carpathians and the Balkan Peninsula, and Asia Minor (Averyanov 1990). Moreover, the highest phylogenetic and genetic diversity, as well as a high degree of endemism within this genus, exists in the Caucasus and Mediterranean regions (Pillon et al. 2006). According to the most recent data, the genus *Dactylorhiza* includes 45 species and 41 subspecies, with 14 species and 10 subspecies recorded in countries of the Balkan Peninsula (WCSP 2020).

Dactylorhiza is a taxonomically extremely complex genus, consisting of seven groups in Europe, North Africa, and the Middle East, viz., the *D. iberica*, *D. sambucina*, *D. incarnata*, *D. majalis*, *D. traunsteineri*, *D. maculata*, and *D. praetermissa* groups (Delforge 2016). In Serbia, taxa from four groups (the *D. sambucina*, *D. incarnata*, *D. maculata*, and *D. majalis* groups) have been recorded (Figure 4). Among the most complex of them are the *D. maculata* and *D. majalis* groups, which require detailed taxonomic and phylogeographic research in the Balkans. Due to its complicated taxonomy, *D. cordigera* requires detailed revision throughout Southeast Europe. Detailed studies on the *Dactylorhiza maculata* group have been carried out in Scandinavia (Ståhlberg and Hedrén 2008; Ståhlberg 2009). However, this group has not been well studied in Serbia. Recent field research (Djordjević 2018) revealed great morphological variability of *Dactylorhiza maculata*, so it is assumed that a fairly large number of "microtaxa" are present, which can be determined by detailed morphological and molecular analyses. According to Averyanov (1990), there are four sections (*Dactylorhiza, Aristatae, Sambucinae*, and *Iberanthus*) and several subsections and aggregates of this genus. To be precise, the section *Dactylorhiza* is the largest one, numbering 64 primitive species in four subsections, viz., *Dactylorhiza* (containing 26 taxa within the aggregates *D. incarnata*, *D. olocheilos*, *D. sanasunitensis*, and *D. salina*), *Maculatae* (containing 18 taxa within the aggregates *D. saccifera*, *D. fuchsii*, and *D. maculata*), *Latifoliae* (containing 16 taxa within the aggregates *D. majalis*, *D. integrata*, *D. cordigera*, and *D. traunsteineri*), and *Sesquipedales* (containing four species within the aggregate *D. sesquipedalis*) (Averyanov 1990). The section *Aristatae* contains the single species *D. aristata*, the section *Iberanthus* is represented by the species *D. iberica*, whereas the section *Sambucinae* numbers nine taxa within the aggregates *D. sambucina* and *D. sulphurea* (Averyanov 1990). *Dactylorhiza* species in Serbia are divided into two sections – the section *Dactylorhiza* (with the subsection *Dactylorhiza* and aggregate *D. incarnata*; the subsection *Maculatae* and aggregates *D. saccifera*, *D. fuchsii*, and *D. maculata*; and the subsection *Latifoliae* and aggregates *D. majalis* and *D. cordigera*); and the section *Sambucinae* (with the aggregates *D. sambucina* and *D. sulphurea*) (Figure 4).

Recent molecular studies have shown that *Coeloglossum viride* belongs to the genus *Dactylorhiza*, which has been given a new taxonomic status – *Dactylorhiza viridis* (L.) R.M. Bateman, Pridgeon & M.W.Chase (Bateman et al. 1997; Pridgeon et al. 1997, 2001). Accordingly, the taxon *D. viridis* has become widely accepted in the floristic literature (Fay et al. 2015; Bateman and Rudall 2018; WCSP 2020). However, Devos et al. (2006) showed that the genus *Coeloglossum* represents a distinct phylogenetic lineage among all representatives of *Dactylorhiza*.

The genus *Anacamptis* (Figure 5) includes 11 species and seven subspecies (WCSP 2020) distributed throughout Europe – to Britain and southern Scandinavia in the north, to northern and northwestern Africa in the south, and to western Asia, Saudi Arabia, and Yemen in the east (Kretzschmar et al. 2007; Tsiftsis and Antonopoulos 2017). This genus ranks second in terms of the number of taxa within the subfamily Orchidoideae in Serbia (Figure 3).

Figure 5. Some representatives of the genus *Anacamptis* in Serbia: a) *Anacamptis coriophora* subsp. *coriophora*, b) *Anacamptis morio* subsp. *morio*, c) *Anacamptis papilionacea*, d) *Anacamptis palustris* subsp. *palustris*, e) *Anacamptis palustris* subsp. *elegans*, f) *Anacamptis pyramidalis* (a, b, f: photos taken from 2014 to 2017 by V. Djordjević; c: photo taken in 2020 by A. Maksimović; d: photo taken in 2020 by I. Stevanoski; e: photo taken in 2019 by E. Kabaš).

The genus was established by the French botanist Louis Claude Marie Richard in 1817 on the basis of *Anacamptis pyramidalis* and was for a long time a monospecific genus (Wood and Ramsay 2004). However, based on the results of molecular systematic studies, *Anacamptis* has been expanded to include taxa previously classified within the genus *Orchis* s.l. (sections *Boryae*, *Coriophorae*, *Laxiflorae*, *Moriones*, *Papilionaceae*, and *Saccatae*) that have a diploid chromosome number – 2n = 36 (2n= 32 in the case of *A. papilionacea*) (Pridgeon et al.

1997; Bateman et al. 1997, 2003; Tichy and Del Prete 2001). Knowledge about the distribution of *Anacamptis laxiflora*, *A. palustris* subsp. *palustris*, and *A. palustris* subsp. *elegans* in Serbia is insufficient (Niketić et al. 2018), as these taxa are presented in the literature with species rank for the flora of Serbia, i.e., as *Orchis laxiflora* Lam. (Diklić 1976). Consequently, the published data on the distribution of these taxa cannot be considered with any degree of certainty, and there is therefore a need for further confirmation of the presence of these taxa in Serbia. Recent research has confirmed that *A. laxiflora* grows in southern Serbia, near Preševo (Niketić et al. 2018) and in the Kosovo region (Niketić et al. 2020).

Figure 6. Some representatives of the genus *Orchis* in Serbia: subgenus *Orchis* s.s. – a) *Orchis militaris*, b) *Orchis purpurea*, c) *Orchis simia*; subgenus *Masculae* – d) *Orchis mascula* subsp. *speciosa*, e) *Orchis pallens*, f) *O. spitzelii* (a-e: photos taken from 2013 to 2015 by V. Djordjević; f: photo taken in 2018 by S. Tsiftsis).

Little is known about the distribution of *Anacamptis morio* subsp. *caucasica*, due to confusion with the typical subspecies (*Anacamptis morio* subsp. *morio*). Moreover, the presence of *Anacamptis coriophora* subsp. *fragrans* is unconfirmed, considering that it is a typically Mediterranean taxon whose occurrence in Serbia has not been confirmed by field research so far.

The genus *Orchis* includes 23 species and 10 subspecies (WCSP 2020). Although the center of diversity of this genus is in the Mediterranean region, where the greatest number of taxa have been recorded, *Orchis* species are distributed throughout Europe and temperate Asia and North Africa, west to the Azores, Madeira and the Canary Islands, east to the Caucasus, and north to Scandinavia (Delforge 2006). Recent molecular analyses have determined the polyphyletic status of the genus *Orchis* s.l. (Pridgeon et al. 1997, 2001; Bateman et al. 1997, 2003), and many species have been transferred to the genera *Anacamptis* and *Neotinea*. According to Tyteca et al. (2012), the taxa of the genus *Orchis* are divided into two subgenera – *Orchis* s.s. and *Masculae*. Among the orchids of Serbia, the subgenus *Orchis* s.s. includes *O. militaris*, *O. purpurea*, and *O. simia*, while the subgenus *Masculae* includes *O. mascula*, *O. pallens*, *O. provincialis*, and *O. spitzelii* (Figure 6). The presence of *O. italica* Poir. in Serbia is questionable, considering that its finding in Serbia has not been confirmed and is considered negative, i.e., false. No other typically Mediterranean representatives of this genus have been found in Serbia: *O. quadripunctata* Cirillo ex Ten., *O. pauciflora* Ten., *O. anthropophora* (L.) All., and *O. punctulata* Steven ex Lindl. In addition, the distribution of *Orchis mascula* subsp. *mascula* and that of *O. mascula* subsp. *speciosa* is poorly known, as they are mostly cited in the literature in the species rank for the flora of Serbia. Although Kretzschmar et al. (2007) indicated that both subspecies occur in the central Balkans, recent field work and examination of available herbarium material have only confirmed the occurrence of the subspecies *speciosa*, so the possible occurrence of the typical subspecies (*mascula*) requires further investigation.

The genus *Ophrys* is represented in Serbia by a relatively small number of species (Figure 7), which is not surprising considering that the center of diversity of this genus is in the Mediterranean region (Delforge 2006; Pedersen and Faurholdt 2007), and the fact that the typical Mediterranean climate and vegetation are not present in Serbia. The genus includes 90 species and subspecies (WCSP 2020), which are distributed from the Canary Islands in the west to the Caspian Sea in the east, and from southern Scandinavia in the north and North Africa (northern Morocco, Algeria, Tunisia, and Libya) and southwest Asia in the south (Delforge 2006; Pedersen and Faurholdt 2007). The genus is divided into two sections – *Pseudophrys* Godfery and *Euophrys* Godfery (nom. nud.), based on structural differences of the lip, i.e., according to the position occupied by the pollinating insect during pseudocopulation on the lip (Antonopoulos and Tsiftsis 2017). According to Delforge (2006), 12 groups are represented within the section *Pseudophrys*, while the section *Euophrys* is represented by 20 groups. Most representatives of the genus *Ophrys* recorded in Serbia belong to the section *Euophrys* and are classified into five groups (*O. insectifera*, *O. apifera*, *O. sphegodes*, *O. fuciflora*, and *O. oestrifera*) (Figure 7), while the section *Pseudophrys* is represented by only one group (*Ophrys lutea*).

Figure 7. Some representatives of the genus *Ophrys* in Serbia: a) *Ophrys apifera*, b) *Ophrys insectifera* subsp. *insectifera*, c) *Ophrys scolopax* subsp. *cornuta* (photo taken from 2014 to 2017 by V. Djordjević).

Species of the genus *Gymnadenia* (Figure 8) are distributed in Europe and Asia, east to western and central Asia, China, and Japan, and south to the Himalayas (Delforge 2006; Tsiftsis and Antonopoulos 2017). According to WCSP (2020), this genus is represented by 26 species and subspecies. A member of this genus, *Gymnadenia frivaldii* (Figure 8b) was previously classified as belonging either to the genus *Pseudorchis* or to the genus *Leucorchis* (Bateman et al. 2006). According to studies based on the results of molecular analyses, the genus *Gymnadenia* also includes taxa from the genus *Nigritella Rich.* (Teppner and Klein 1998; Bateman et al. 2003). However, recent research has restored the genus *Nigritella* as distinct (Hedrén et al. 2018b). The center of diversification of the genus *Nigritella* is in the Alps (Hedrén et al. 2000). The species *Nigritella* (*Gymnadenia*) *nigra* in the botanical literature has been mentioned as occurring in Serbia, and it is now known to be an endemic species of Scandinavia (Delforge 2006). Field research and revision of herbarium material have revealed the presence in Serbia of *Nigritella* (*Gymnadenia*) *rhellicani* (Figure 8d), which was also cited as a subspecies – *Nigritella nigra* subsp. *rhellicani* (Teppner & E. Klein) H. Baumann, Künkele & R. Lorenz. However, recent studies have confirmed taxonomic validity of the species *Nigritella rhellicani* Teppner & E. Klein (Hedrén et al. 2018b).

The genus *Neotinea* (Figure 9) includes four species and one subspecies (Chase et al. 2015; WCSP 2020), distributed in Europe, Asia Minor, the Caucasus, and northwestern regions of North Africa (Kretzschmar et al. 2007). The taxonomic status of the species *N. tridentata, N. ustulata*, and *N. lactea* was recently determined by molecular analyses (Bateman et al. 2003). Previously, these species were classified within the genus *Orchis* (Moore 1980). In addition, the authors of recent studies have described two varieties of *N. ustulata*: var. *ustulata* (early-flowering) and var. *aestivalis* (late-flowering) (Tali et al. 2004, 2006; Haraštová-Sobotková et al. 2005). It is important to mention the presence of the hybrid *Neotinea* × *dietrichiana* (*N. tridentata* × *N. ustulata*), which was recorded not only for the first time in western

Serbia, but also for the first time throughout the entire territory of Serbia (Djordjević et al. 2012).

Figure 8. Representatives of the genera *Gymnadenia* and *Nigritella* in Serbia: a) *Gymnadenia conopsea*, b) *Gymnadenia frivaldii*, c) *Gymnadenia odoratissima*, d) *Nigritella rhellicani* (a, b, d: photos taken from 2006 to 2014 by V. Djordjević; c: photo taken in 2017 by M. Bobocea).

The genus *Himantoglossum* comprises 12 taxa, i.e., 10 species and two subspecies, distributed in North Africa, Europe, Asia Minor, the Caucasus, Israel, and southwestern Iran (Tsiftsis and Antonopoulos 2017; WCSP 2020). Species formerly included in the genera *Barlia* Parl. and *Comperia* K. Koch are also placed in this genus, since molecular research has established their relationship (Bateman et al. 2003). Five taxa of this genus are represented on the Balkan Peninsula: *H. adriaticum* H. Baumann, *H. calcaratum* (Beck) Schltr. subsp. *calcaratum* (Figure 10), *H. calcaratum* subsp. *jankae* (Somlyay, Kreutz & Óvári) R.M. Bateman, Molnar & Sramkó, *H. comperianum* (Steven) P. Delforge, and *H. robertianum* (Loisel.) P. Delforge (Tsiftsis 2016; Bateman et al. 2017a; Tsiftsis and Antonopoulos 2017; WCSP 2020).

Figure 9. Representatives of the genus *Neotinea* in Serbia: a) *Neotinea tridentata*, b) *Neotinea ustulata* (photos taken in 2013 and 2014 by Djordjević).

According to Bateman et al. (2017a), the genus *Himantoglossum* is divided into three subgenera (*Himantoglossum*, *Barlia*, and *Comperia*), and there are three sections (*Hircinum, Caprinum*, and *Formosum*) within the subgenus *Himantoglossum*. The only representatives of this genus in Serbia (*H. calcaratum* subsp. *calcaratum* and *H. calcaratum* subsp. *rumelicum*) belong to the section *Caprinum*. *Himantoglossum calcaratum* subsp. *jankae* (Molnár et al. 2012) was cited in the floristic literature in Serbia as *H. hircinum* Spreng. subsp. *caprinum* (M. B.) Hayek var. *caprinum* (Diklić 1976). This taxon was renamed *H. jankae* Somlyay,

Kreutz & Óvári (Molnár et al. 2012) and was later granted subspecies status – *H. calcaratum* subsp. *jankae* (Bateman et al. 2017a). However, the newly established nomenclatural combination *H. calcaratum* subsp. *rumelicum* was officially included in the existing checklist of the orchid flora in Serbia (Djordjević et al. 2018; Niketić et al. 2018).

Figure 10. *Himantoglossum calcaratum* subsp. *calcaratum*, Serbia (photos taken in 2013 and 2014 by V. Djordjević).

The genus *Platanthera* includes 146 species and subspecies, distributed in Europe, North Africa, Asia, and Central and North America (WCSP 2020). The genus is divided into five subgenera (Efimov 2016). Although only two species (*P. bifolia* and *P. chlorantha*) (Figures 11a and 11b) have been identified within the subgenus *Platanthera* and the section *Platanthera* in Serbia, it is considered that a number of "microtaxa" are present, in view of the fact that the species inhabit both herbaceous and forest vegetation types, while considerable morphological variability of populations has been observed. Three subspecies of *P. bifolia* have been described to date: *P. bifolia* subsp. *bifolia* (a widespread subspecies), *P. bifolia* subsp. *osca* R. Lorenz,

Romolini, V.A. Romano & Soca (Italy), and *P. bifolia* subsp. *subalpina* Brügger (Austria, Germany, Switzerland, Italy) (WCSP 2020). Although *Platanthera chlorantha* is accepted as a species-level taxon (WCSP 2020), some authors distinguish a typical subspecies (*chlorantha*) and *P. chlorantha* subsp. *holmboei* (H.Lindb.) J.J. Wood growing in Cyprus and Greece (Meikle 1985; Tsiftsis and Antonopoulos 2017).

The genus *Pseudorchis* is represented by three species and subspecies – *P. albida* (L.) Á. Löve & D. Löve subsp. *albida*, *P. albida* subsp. *tricuspis* (Beck) E. Klein, and *P. straminea* (Fernald) Soják (Bateman et al. 2017b; WCSP 2020). These taxa are widely distributed in the boreal and temperate regions of Europe and Asia (WCSP 2020). Taxa of this genus were classified within the genera *Gymnadenia* and *Leucorchis*, but recent molecular studies based on ITS (internal transcribed spacer) sequences of nuclear ribosomal DNA have shown that the genus *Pseudorchis* is close to the genus *Platanthera*, so that both genera are within the same clade: *Pseudorchis-Amerorchis-Galearis-Neolindleya-Platanthera* (Bateman et al. 2003; Jersáková et al. 2011). The Eurasian genus *Herminium* includes 49 species and subspecies, of which only the species *H. monorchis* (Figure 11c) occurs in Europe (WCSP 2020). At the same time, *H. monorchis* is the only representative of this genus in Serbia.

In the orchid flora of Serbia, there is one species of the genus *Traunsteinera* – *T. globosa* (Figure 11d). In addition to that species, this genus also includes the species *T. sphaerica* (M. Bieb.) Schltr., which is distributed in the Caucasus and Anatolia (Delforge 2006; WCSP 2020). Recent molecular studies have shown that this genus is most closely related to the genus *Chamorchis*, with which it forms the clade *Traunsteinera-Chamorchis* (Bateman et al. 2003).

Within the subfamily Orchidoideae, the orchids of Serbia are also represented by members of the tribe Cranichideae, with two (Goodyerinae and Spiranthinae) out of a total of eight subtribes (Chase et al. 2015). The subtribe Goodyerinae includes orchids from 34 genera, distributed mainly in tropical and subtropical areas of America, Africa, and Asia (Bernardos et al. 2007; Chase et al. 2015). Among the genera of this subtribe, *Goodyera* stands out, with 98 species (Chase et al. 2015). This genus is represented by only one species (*Goodyera repens*) in Europe, as in Serbia (Figure 11e). The subtribe Spiranthinae includes species from 41 genera (Chase et al. 2015), distributed mainly in tropical areas of America (Bernardos et al. 2007). This subtribe includes the cosmopolitan genus *Spiranthes*, with 34 species, three of which are present in Europe – *Spiranthes spiralis* (L.) Chevall. (Figure 11f), *S. aestivalis* (Poir.) Rich., and *S. romanzoffiana* Cham. Two species of this genus (*S. spiralis* and *S. aestivalis*) are present in the flora of Serbia (Diklić 1976; Djordjević et al. 2018).

Figure 11. Representatives of the genera *Platanthera, Herminium, Traunsteinera, Goodyera*, and *Spiranthes* in Serbia: a) *Platanthera bifolia*, b) *Platanthera chlorantha*, c) *Herminium monorchis*, d) *Traunsteinera globosa*, e) *Goodyera repens*, f) *Spiranthes spiralis* (a, b, d, e: photos taken from 2006 to 2020 by V. Djordjević; c: photo taken in 2018 by M. Bobocea; f: photo taken in 2019 by I. Stevanoski).

The subfamily Epidendroideae is the largest subfamily of orchids and includes about 22,000 species classified into 535 genera, distributed mainly in the tropics of South America and Asia (Chase et al. 2017). The representatives of this subfamily are for the most epiphytes (Chase et al. 2017). In Europe, representatives from the tribes Neottieae, Nervilieae, Malaxideae and Epidendreae are present (Chase et al. 2015). This subfamily consists of a total of 16 tribes (Chase et al. 2015), members of three of them being present in Serbia (Neottieae, Nervilieae, and Epidendreae) (Figure 2).

The tribe *Neottieae* includes 178 species from six genera, distributed mainly in the Northern Hemisphere, and their southern distribution limits coinciding with subtropical and tropical regions of America, Africa, and Asia (Chase et al. 2015). Orchids of four genera within this tribe (*Cephalanthera*, *Epipactis*, *Limodorum*, and *Neottia*) have been found in Serbia. The large number of species within the genus *Epipactis* in Serbia (Figure 3) is understandable given the significant presence of forest ecosystems, especially beech forests, which are the main habitats of the genus (Delforge 2006). This genus comprises a total of 79 taxa (58 species and 21 subspecies; WCSP 2020), distributed mainly in Europe and Asia, with a small number of representatives in North Africa and North America (Pridgeon et al. 2005; Pedersen et al. 2013). The main centers of diversity of the genus *Epipactis* are Western Europe, Central Europe, and Western Asia (Efimov 2008). In addition, recent studies have shown that the Apennine and Balkan Peninsulas are also important centers of diversity of this genus (Tranchida-Lombardo et al. 2011). On the Balkan Peninsula, a significant number of *Epipactis* taxa have been recorded in Greece (21 taxa) (Tsiftsis and Antonopoulos 2017), Croatia (17 taxa) (Nikolić 2019), Slovenia (15 taxa) (Dolinar 2015), and Bulgaria (11 taxa) (Assyov and Petrova 2012). A total of 66 species and subspecies of the genus *Epipactis* grow in Europe, North Africa, and the Middle East, and they are divided into two sections (*Arthrochilium* and *Euepipactis*) and nine groups (Delforge 2006). Representatives from both sections and five groups (*E. atrorubens*, *E. purpurata*, *E. helleborine*, *E. leptochila*, and *E. palustris*) occur in Serbia (Figure 12).

Figure 12. Some representatives of the genus *Epipactis* in Serbia: a) *Epipactis atrorubens*, b) *Epipactis helleborine* subsp. *helleborine*, c) *Epipactis palustris*, d) *Epipactis purpurata* (photos taken from 2013 to 2016 by V. Djordjević).

The genus *Cephalanthera* (Figure 13) includes 19 species, distributed in Europe, North Africa, the Middle East, the Caucasus, West Asia, China, Indochina, and Japan (Chase et al. 2015; Tsiftsis and Antonopoulos 2017; WCSP 2020). *Cephalanthera austiniae* (A. Gray) A. Heller is the only species of this genus growing in North America

(WCSP 2020). Out of a total of five species found in Europe, three species are present in the flora of Serbia. The remaining two taxa are *C. cucullata* Boiss. & Heldr. (an endemic species on the island of Crete) and *C. epipactoides* Fisch. & C.A. Mey. (distributed in Greece, Bulgaria, and Turkey) (Tsiftsis and Antonopoulos 2017).

The genus *Neottia* is represented by 72 species (WCSP 2020), distributed mainly in temperate areas of Europe, Asia, and North America, and less frequently in tropical areas of Asia (Pridgeon et al. 2005; Delforge 2006). Three *Neottia* species are present in the orchid flora of Serbia. Recently, species from the genus *Listera* R. Br. were also included in this genus, as their relationship was established by molecular studies. Representatives of the genus *Limodorum* are distributed in Southern and Central Europe, Southwest Asia, and North Africa (Tsiftsis and Antonopoulos 2017). This genus is represented by three species (WCSP 2020), of which only one (*L. abortivum*) grows in Serbia.

Figure 13. Representatives of the genus *Cephalanthera* in Serbia: a) *Cephalanthera damasonium*, b) *Cephalanthera longifolia*, c) *Cephalanthera rubra* (photos taken from 2017 to 2020 by V. Djordjević).

The tribe Nervilieae includes two subtribes (Nerviliinae and Epipogiinae) and three genera (*Nervilia*, *Epipogium*, and *Stereosandra*) (Chase et al. 2015). Species from the genera *Nervilia* and *Stereosandra* are mainly distributed in Indonesia and Malaysia (Bernardos et al. 2007), while *Epipogium* species are distributed in Europe and Asia (WCSP

2020). *Epipogium aphyllum* is the only representative of the tribe Nervilieae in Serbia.

The tribe Epidendreae includes six subtribes: Bletiinae, Laeliinae, Pleurothallidinae, Ponerinae, Calypsoinae, and Agrostophyllinae (Chase et al. 2015). The subtribe Calypsoinae includes species from 13 genera, distributed mostly in the Northern Hemisphere (Bernardos et al. 2007; Chase et al. 2015). Among the genera, the genus *Corallorhiza* includes 11 species, of which only *C. trifida* occurs in Europe. It is also the only species from the tribe Epidendreae present in Serbia.

The subfamily Cypripedioieae is represented by 169 species from five genera (*Cypripedium*, *Mexipedium*, *Paphiopedilum*, *Phragmipedium*, and *Selenipedium*), distributed in tropical and north temperate regions of America and Asia, as well as in Europe (Chase et al. 2017). Representatives of this subfamily are mainly terrestrial herbaceous orchids characterized by the presence of two or three fertile anthers (Dressler 1981; Chase et al. 2017). The genus *Cypripedium* includes 52 species and subspecies (WCSP 2020), with only one species (*C. calceolus*) occurring in Serbia.

LIFE FORMS

Within the family Orchidaceae, a great diversity of life forms and life histories has been found, both in terms of morpho-anatomical and physiological adaptations, functional traits, and survival during unfavorable periods of the year, and in terms of mycorrhizal associations and pollination systems (Whigham and Willems 2003; Swarts and Dixon 2009; Taylor et al. 2021). With respect to orchid life forms, epiphytic orchids of tropical and subtropical areas constitute the majority of representatives, while terrestrial orchids account for a quarter of the total number of taxa (Dressler 1981). Most epiphytic orchids have succulent leaves and aerial roots covered with a layer of dead cells (*velamen radicum*) that have the ability to absorb water rapidly (Zotz and Winkler 2013). Some epiphytic orchids have CAM photosynthesis to conserve water (Silvera et al. 2009), while some have pseudobulbs that store water (Taylor et al. 2021). Thus, these orchids survive in forest canopies, where water availability is an important limiting factor.

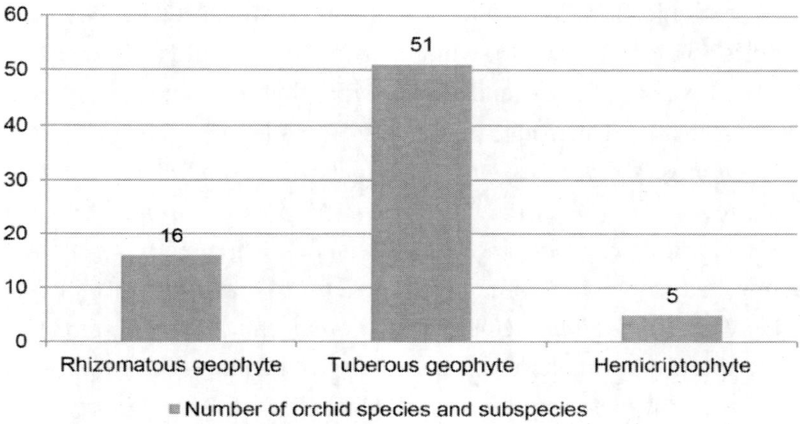

Figure 14. Structure of orchid life forms in Serbia according to WCSP (2020).

Most representatives of the family Orchidaceae of the temperate climate zone, and thus also of Serbia (Table 1), represent terrestrial orchids that display characteristics of the geophyte life form (life form I) (Raunkiaer 1934; Ellenberg and Mueller-Dambois 1967; Taylor et al. 2021). This life form is that of an ecological group of plants whose above-ground herbaceous organs die completely during the unfavorable climatic season, while the buds that give rise to new shoots the following year are located in underground organs (bulbs, corms, tubers, or rhizomes) in the soil. The characteristics of geophytes are associated with the storage of carbohydrates, which allows for their survival during long winters or dry and warm summers (Raunkiaer 1934; Taylor et al. 2021). The geophyte life form is mainly found in Mediterranean climates and in temperate zones where unfavorable climatic conditions such as winter frost and hot and dry summers prevail (Taylor et al. 2021). However, there are also non-geophytes among terrestrial orchids, i.e., orchids belonging to the category of chamaephytes (e.g., *Dendrobium vannouhuysii*) and that of hemicryptophytes, which have only above-ground buds, allowing the plants to survive the coldest winter months under the protection of snow cover (Ellenberg and Mueller-Dambois 1967; Taylor et al. 2021). The structure of life forms of the orchids of Serbia is shown in Figure 14.

The orchid flora in Serbia is dominated by tuberous geophytes, which comprise 51 taxa (70.83%), while rhizomatous geophytes are represented by 16 taxa (22.22%) and hemicryptophytes comprise five orchid taxa (7.95%). Tuberous geophytes include taxa from 14 genera (*Anacamptis*, *Coeloglossum*, *Dactylorhiza*, *Gymnadenia*, *Herminium*, *Himantoglossum*, *Neotinea*, *Nigritella*, *Ophrys*, *Orchis*, *Platanthera*, *Pseudorchis*, *Spiranthes*, and *Traunsteinera*). Rhizomatous geophytes are represented by taxa from seven genera (*Corallorhiza*, *Cypripedium*, *Epipactis*, *Epipogium*, *Goodyera*, *Limodorum*, and *Neottia*), while hemicriptophytes include taxa from two genera (*Cephalanthera* and *Neottia*). However, on the World Checklist of Kew Gardens (WCSP 2020), three taxa were reported as hemicryptophytes or rhizome geophytes (*Epipactis leptochila* subsp. *neglecta*, *E. palustris*, and *E. purpurata*), while two taxa were reported as hemicryptophytes or tuber geophytes (*Spiranthes aestivalis* and *S. spiralis*). In our analyses, these taxa were considered as geophytes (Table 1; life form I).

Among the orchids with rhizomes, there are four holomycotrophic taxa (*Corallorhiza trifida*, *Epipogium aphyllum*, *Limodorum abortivum*, and *Neottia nidus-avis*), which represent species that are obligate mycotrophs (Figure 15). This life form has often been incorrectly described as saprophytic. Chlorophyll is completely absent in the species *Epipogium aphyllum* and *Neottia nidus-avis*, while a small amount of chlorophyll is present in *Limodorum abortivum* and *Corallorhiza trifida* (Delforge 2006). *Neottia nidus-avis* is specialized on fungi of the family Sebacinaceae, which are known for forming ectomycorrhiza with trees (McKendrick et al. 2002; Selosse et al. 2002). Recent studies have shown that orchids from the tribe Neottieae, which have the ability to photosynthesize, also use ectomycorrhizal relationships with other plants to replenish their own carbon stocks (Girlanda et al. 2006; Waterman and Bidartondo 2008).

Figure 15. Holomycotrophic orchid taxa in Serbia: a) *Corallorhiza trifida*, b) *Epipogium aphyllum*, c) *Limodorum abortivum*, d) *Neottia nidus-avis* (a: photo taken in 2007 by S. Tsiftsis; b-d: photos taken from 2014 to 2019 by V. Djordjević).

This combined feeding, i.e., partial mycoheterotrophy or myxotrophy, has been found in *Epipactis microphylla*, whose individuals form communities with ectomycorrhizal fungi of the genus *Tuber*, which in turn form ectomycorrhizal associations with trees (Selosse et al. 2004). The photosynthetic capacity of *Limodorum abortivum* is insufficient and replenishment of carbon reserves occurs through association with ectomycorrhizal species of the genus *Russula* (Girlanda et al. 2006). Until recently, partial mycoheterotrophy in orchids growing in forest habitats was thought to occur due to light deficiency. However, recent studies show that partial mycoheterotrophy also occurs in orchids growing in grassland habitats under a favorable light regime (Schiebold et al. 2018).The evolutionary development of underground organs in orchids proceeded from rhizomes to tubers. It follows that the most primitive representatives of European orchids are species from the genera *Cephalanthera* and *Epipactis*, which have short rhizomes (Figure 16a) (Averyanov 1990). The next stage in the evolution of terrestrial orchids was the development of "intermediate orchids" (in evolutionary history intermediate between rhizomatous and ovoid-tuberous orchids) (Averyanov 1990; Dressler 1981).

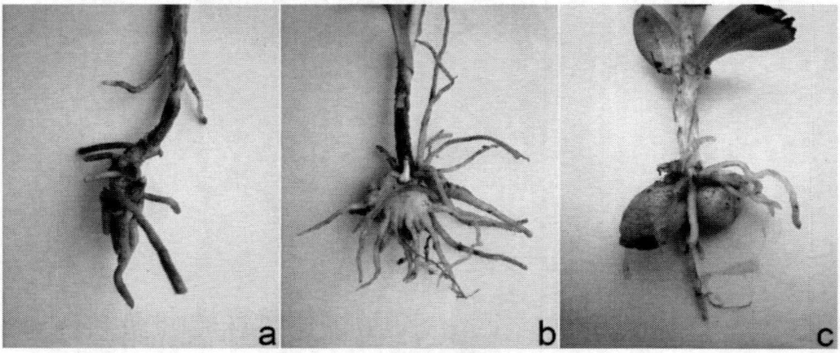

Figure 16. Different types of orchid root systems: (a) rhizome – *Cephalanthera damasonium*, (b) palmate tubers ("intermediate") – *Gymnadenia conopsea* and (c) ovoid tubers – *Neotinea ustulata* (photos taken in 2020 by V. Djordjević).

The appearance of "intermediate orchids" with palmately lobed (finger-like) and fusiform tubers has been linked with the Alpine orogeny, i.e., the emergence of lower temperature mountain habitats at the end of the Paleogene or early Neogene (Averyanov 1990). The same author claims that these orchids significantly expanded their range as a result of cooling at the end of the Neogene and in the Pleistocene. Under such conditions, the species could occupy areas with plains, where degradation of the thermophilic Tertiary flora took place. The first formation of finger-like tuberoids occurred in relatively primitive orchids (*Dactylorhiza incarnata* and *D. fuchsii*) and in related genera (*Pseudorchis*, *Coeloglossum*, and *Gymnadenia*) (Figure 16b), after which increasing specialization led to cylindrical tuberoids in advanced species such as *Dactylorhiza sulphurea* and *D. sambucina* (Averyanov 1990). Ovoid and spindle-shaped tubers represent the final stage in evolution of the underground organs of terrestrial orchids and enable orchids to survive in dry and warm conditions (Figure 16c) (Averyanov 1990; Tatarenko 2007). The structure of life forms of orchids of Serbia according to this classification is shown in Figure 17 and Table 1 (life form II). The root system of some orchids in Serbia on whose basis the classification of life forms was made is presented in Figure 16.

The orchid flora in Serbia is dominated by orchids with ovoid and spindle-shaped tubers (Figure 16c; Table 1). This group includes 33 orchid species (45.83%) from eight genera (*Anacamptis*, *Herminium*, *Himantoglossum*, *Neotinea*, *Ophrys*, *Orchis*, *Spiranthes,* and *Traunsteinera*; Figure 16c). Orchids with rhizomes include 21 orchid species and subspecies (29.17%) from eight genera (*Cephalanthera*, *Corallorhiza*, *Cypripedium*, *Epipactis*, *Epipogium*, *Goodyera*, *Limodorum*, and *Neottia*; Figure 16a), while representatives of "intermediate orchids" with palmately lobed and fusiform tubers are represented by 18 species and subspecies (25.00%) from six genera (*Coeloglossum*, *Dactylorhiza*, *Gymnadenia*, *Nigritella*, *Platanthera*, and *Pseudorchis*; Figure 16b). Among these orchids, *Pseudorchis albida* is the most primitive species, species of the genus *Platanthera* have fusiform tubers, and taxa of the genera *Coeloglossum*, *Dactylorhiza*, *Nigritella*, and *Gymnadenia* (Figure 16b) are characterized by palmately lobed (finger-like) tubers. The percentage of specific life forms in Serbia differs significantly from the representation of life forms in Greece, where orchids with ovoid tubers completely dominate, with 138 taxa (71.13% of the total number of taxa) (Tsiftsis and Antonopoulos 2017; Antonopoulos and Tsiftsis 2017). They are followed by orchids with rhizomes (35 taxa or 18.04%), while orchids with palmately lobed and fusiform tubers are the least represented (21 taxa or 10.83%) (Tsiftsis and Antonopoulos 2017; Antonopoulos and Tsiftsis 2017; Tsiftsis and Tsiripidis 2020). These data are consistent with the hypothesis that ovoid tubers are best adapted to warm and dry conditions (Averyanov 1990;

Tatarenko 2007), which are characteristic of the Mediterranean area of Greece. The higher percentage of orchids with rhizomes and palmately lobed and fusiform tubers in Serbia compared to Greece is explained by the greater influence of humid and continental climate, as well as by the significant presence of forests, wet meadows, fens, and marshes in Serbia, where these orchids are most abundant (Averyanov 1990; Pillon et al. 2006). In Serbia, orchids with rhizomes were recorded mainly in forest ecosystems (Djordjević et al. 2020a, 2020b), while orchids with palmately lobed and fusiform tubers were found for the most part in mesophilous and hygrophilous grasslands and fen habitats (Djordjević et al. 2016b), which is consistent with results of previous studies on orchids in Europe (Buttler 1991; Delforge 2006).

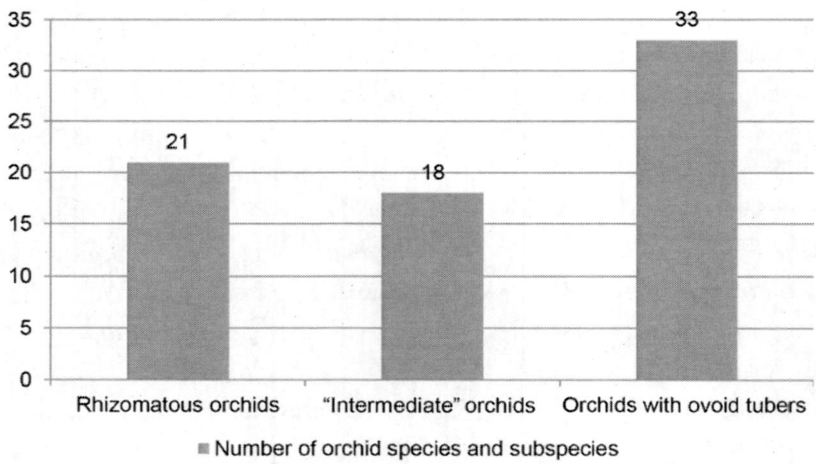

Figure 17. Structure of life forms of orchids of Serbia according to the classification proposed by Dressler (1981), Averyanov (1990), Tatarenko (2007), Tsiftsis et al. (2019b), and Štípková et al. (2021).

Most orchids in Serbia bloom in the spring-summer period (35 taxa), followed by orchids that bloom in spring (25 taxa) and summer (10 taxa), while the smallest number of orchids bloom in the summer-autumn period (two taxa) (Table 1). Orchids in Serbia bloom from the end of March and beginning of April until the last week of September.

Table 1. Overview of life forms I (TG – tuberous geophyte, RG – rhizomatous geophyte, hRG – holomycotrophic rhizomatous geophyte, H – hemicryptophyte); life forms II (RH – rhizomatous orchids, I – "intermediate orchids", T – tuberous orchids); and flowering time (v – spring, a – summer, aut – autumn) of orchids in Serbia

Orchid taxon	Life form I	Life form II	Flowering Period
Anacamptis coriophora subsp. *coriophora*	TG	T	v-a: mid-May – early July
Anacamptis coriophora subsp. *fragrans*	TG	T	v: April – June
Anacamptis laxiflora	TG	T	v: May – June
Anacamptis morio subsp. *morio*	TG	T	v: mid-April – early June
Anacamptis morio subsp. *caucasica*	TG	T	v: mid-April – early June
Anacamptis palustris subsp. *palustris*	TG	T	v: May – early July
Anacamptis palustris subsp. *elegans*	TG	T	v: May – early July
Anacamptis papilionacea subsp. *papilionacea*	TG	T	v: May – June
Anacamptis pyramidalis	TG	T	v-a: May – July
Cephalanthera damasonium	H	RH	v-a: mid-May – early July
Cephalanthera longifolia	H	RH	v-a: April – June
Cephalanthera rubra	H	RH	v-a: May – July
Coeloglossum viride	TG	I	v-a: May – early July
Corallorhiza trifida	hRG	RH	v-a: mid-May – June
Cypripedium calceolus	RG	RH	v-a: May – early July
Dactylorhiza cordigera subsp. *cordigera*	TG	I	v-a: June – July (– early August)
Dactylorhiza cordigera subsp. *bosniaca*	TG	I	v-a: June – July (– early August)
Dactylorhiza fuchsii subsp. *fuchsii*	TG	I	v-a: June – July
Dactylorhiza incarnata subsp. *incarnata*	TG	I	v-a: May – early July
Dactylorhiza maculata subsp. *maculata*	TG	I	v-a: June – July (– mid-August)
Dactylorhiza maculata subsp. *transsilvanica*	TG	I	v-a: June – July (– mid-August)

Table 1. (Continued)

Orchid taxon	Life form I	Life form II	Flowering Period
Dactylorhiza majalis subsp. *majalis*	TG	I	v-a: May – July
Dactylorhiza romana subsp. *romana*	TG	I	v: March – May
Dactylorhiza saccifera subsp. *saccifera*	TG	I	v-a: mid-May – early August
Dactylorhiza sambucina	TG	I	v: mid-April – June
Epipactis atrorubens	RG	RH	a: June – July (– early August)
Epipactis distans	RG	RH	a: July (– early August)
Epipactis helleborine subsp. *helleborine*	RG	RH	a: late June – mid-August
Epipactis leptochila subsp. *neglecta*	RG	RH	a: early July – early August
Epipactis microphylla	RG	RH	v-a: early June – mid-July
Epipactis muelleri subsp. *muelleri*	RG	RH	a: late-June – July (– early August)
Epipactis palustris	RG	RH	v-a: June – July
Epipactis pontica	RG	RH	a: late July – August
Epipactis purpurata	RG	RH	a: late July – August
Epipactis tallosii subsp. *tallosii*	RG	RH	a: late July – early September
Epipogium aphyllum	hRG	RH	a: mid-July – August
Goodyera repens	RG	RH	a: mid-July – mid-August
Gymnadenia conopsea	TG	I	v-a: mid-May – July
Gymnadenia frivaldii	TG	I	v-a: June – July
Gymnadenia odoratissima	TG	I	v-a: late May – mid-July
Herminium monorchis	TG	T	v-a: May – July
Himantoglossum calcaratum subsp. *calcaratum*	TG	T	v-a: mid-June – early August
Himantoglossum calcaratum subsp. *rumelicum*	TG	T	v-a: mid-June – early August
Limodorum abortivum	hRG	RH	v: mid-May – June
Neotinea lactea	TG	T	v: March – May
Neotinea tridentata subsp. *tridentata*	TG	T	v: late April – June
Neotinea ustulata	TG	T	v-a: late April – July
Neottia cordata	H	RH	v-a: May – June
Neottia nidus-avis	hRG	RH	v-a: late April – June
Neottia ovata	H	RH	v-a: May – July
Nigritella rhellicani	TG	I	v-a: June – July
Ophrys apifera	TG	T	v: May – June
Ophrys fuciflora	TG	T	v: May – June
Ophrys insectifera subsp. *insectifera*	TG	T	v: May – June
Ophrys sicula	TG	T	v: March – May

Orchid taxon	Life form I	Life form II	Flowering Period
Ophrys scolopax subsp. *cornuta*	TG	T	v-a: late April – June (–July)
Ophrys sphegodes subsp. *sphegodes*	TG	T	v: (late March –) April – May
Ophrys sphegodes subsp. *taurica*	TG	T	v: March – April (–May)
Orchis mascula subsp. *mascula*	TG	T	v: mid-April – June
Orchis mascula subsp. *speciosa*	TG	T	v: mid-April – June
Orchis militaris subsp. *militaris*	TG	T	v-a: May – June
Orchis pallens	TG	T	v: mid-April – mid-May
Orchis provincialis	TG	T	v: April – May
Orchis purpurea subsp. *purpurea*	TG	T	v: April – early June
Orchis simia subsp. *simia*	TG	T	v: April – early June
Orchis spitzelii subsp. *spitzelii*	TG	T	v: May – June
Platanthera bifolia	TG	I	v-a: May – early July
Platanthera chlorantha subsp. *chlorantha*	TG	I	v-a: late April – early July
Pseudorchis albida subsp. *albida*	TG	I	v-a: mid-June – mid-July
Spiranthes aestivalis	TG	T	a-aut: May – August (– September)
Spiranthes spiralis	TG	T	a-aut: late August – October
Traunsteinera globosa	TG	T	v-a: late May – July

However, orchids in Mediterranean areas start to bloom earlier, as early as February and March, when many species of the genera *Ophrys* and *Orchis* bloom, as does the species *Himantoglossum robertianum* (Delforge 2006; Tsiftsis and Antonopoulos 2017; Antonopoulos and Tsiftsis 2017). The first orchids to flower in Serbia occur mainly in areas with lower altitudes (*Ophrys sphegodes*, *O. sicula*, *Orchis pallens*, *O. simia*, *O. purpurea*, *Anacamptis morio*, and *Cephalanthera longifolia*), which is explained by the higher spring temperatures in these areas. In Serbia, *Epipactis purpurata*, *E. tallosii*, *Epipogium aphyllum*, and *Spiranthes* species flower last. Most orchids with tubers bloom in Serbia in the spring and spring-summer periods, while orchids with rhizomes, especially ones of the genus *Epipactis*, usually bloom in summer. The flowering periods of certain orchid species in Serbia coincide with their previously established flowering periods in Europe (Buttler 1991; Delforge 2006; Baumann et al. 2006; Tsiftsis and Antonopoulos 2017; Antonopoulos and Tsiftsis 2017).

POLLINATION SYSTEMS

Orchid pollination systems are divided into those that reward pollinators and those that do not. In most cases, the pollinators are insects, while birds are the pollinators of a small number of tropical orchids (van der Pijl and Dodson 1966; Nilsson 1992; Micheneau et al. 2006). Most European orchids are pollinated by bees, wasps, and bumblebees (Hymenoptera), butterflies (Lepidoptera), and insects from the orders Coleoptera and Diptera (van der Pijl and Dodson 1966). In addition to cross-pollination, a significant number of orchids are characterized by self-pollination. The presence of specific pollination systems in the orchid flora of Serbia is shown in Figure 18. According to the pollination systems defined by van der Cingel (1995), Jacquemyn et al. (2005a), Claessens and Kleynen (2011), and Inda et al. (2012), the orchids of Serbia can be classified into three groups: rewarding, deceptive, and self-pollinating orchids (Table 2). Known pollinators of orchid taxa are also presented, along with literature sources in Table 2.

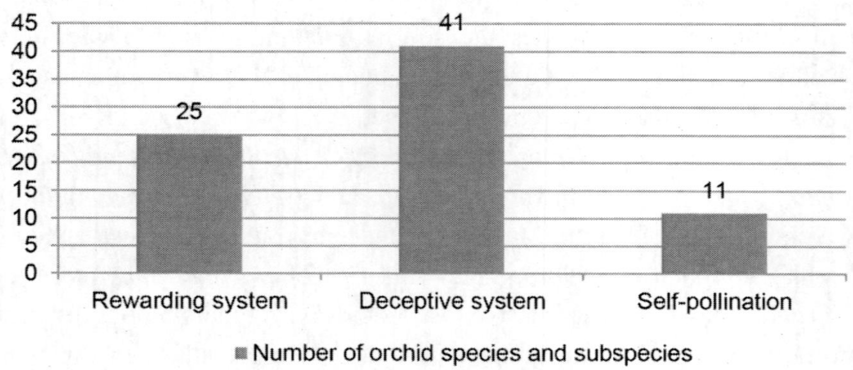

Figure 18. Representation of specific pollination systems in the orchid flora of Serbia.

The deceptive pollination system is present in 41 orchid taxa of Serbia (56.94%) from nine genera (*Anacamptis, Cephalanthera, Cypripedium, Dactylorhiza*, *Himantoglossum*, *Neotinea*, *Ophrys, Orchis,* and *Traunsteinera*) (Table 2). The significant number of orchids in Serbia using the system of deceptive pollination is understandable in view of the fact that about one third of the total number of orchids (between 6500 and 9000 taxa) deceive their pollinators (van der Pijl and Dodson 1966; Dressler 1981; Ackerman 1986; Cozzolino and Widmer 2005; Tremblay et al. 2005). The orchids of Serbia use several mechanisms of pollinator deception. The most common is "generalized food deception", which occurs in orchids of the genera *Anacamptis*, *Cypripedium*, *Dactylorhiza*, *Himantoglossum*, *Neotinea*, and *Orchis* (Jersáková et al. 2006; Vereecken et al. 2010; Fantinato et al. 2017). These orchids exploit the innate foraging behavior of pollinators by emitting signals typical of rewarding plant species (flower color, scent, size, shape, spurs, nectar guides, pollen-like papillae, and inflorescence shape), while not mimicking any particular rewarding plant (Jersáková et al. 2006). Orchids using this mechanism of deception have been found to flower most often in spring, when they take advantage of bees and bumblebees that have just awakened from hibernation (Heinrich 1975). Pollinators may also be insects from a different habitat or those that explore new foraging opportunities because their usual resources have been depleted (Jersáková et al. 2006).

Some orchids of Serbia use Batesian floral mimicry, a mechanism by which orchids achieve pollination through the similarity of their flowers to those of certain rewarding vascular plant species (Jersáková et al. 2006). This type of deception is present in *Traunsteinera globosa*, where the flowers mimic those of species from the genera *Knautia*, *Scabiosa*, *Trifolium*, and *Valeriana* (Jersáková et al. 2016; Chase et al. 2017). According to Kowalkowska et al. (2012), *Anacamptis pyramidalis* also uses this mechanism of deception by imitating flowers of the rewarding orchid *Gymnadenia conopsea*. In addition, the flowers of *Orchis pallens* have been found to imitate *Lathyrus vernus* (L.) Bernh. (Vöth 1982; Dafni 1987), while the flowers of *Orchis provincialis* mimic the flowers of *Lathyrus aureus* (Steven) D. Brândză and *Corydalis cava* subsp. *marschalliana* (Willd.) Hayek (Svolynskiy et al. 2014).

Rendezvous attraction, a mechanism involving the exploitation of male bees' sex drive during flight to find a mate (Jersáková et al. 2016), is present in two orchids from Serbia: *Anacamptis papilionacea* (Vogel 1972) and *Cephalanthera rubra* (Nilsson 1983b). Male bees are deceived by orchids with similar shape, color, or scent as co-flowering rewarding plants (Jersáková et al. 2016). *Cephalanthera rubra* attracts several bee species, especially *Chelostoma rapunculi*, and *C. campanularum* (Hymenoptera, Megachilidae) (Nilsson 1983b; Sedivy et al. 2008). Colorimetric analysis of floral reflection revealed that *C. rubra* mimics *Campanula* species that are regularly visited by bees (Nilsson 1983b; Vereecken et al. 2010). In addition, the middle lobe of the lip of the flower of *C. rubra* has cream-colored wavy wrinkled outer ribs that most likely mimic the reproductive organs of *Campanula* flowers (Vereecken et al. 2010). Flowering of *C. rubra* occurs about two weeks before that of *Campanula* species, when the first males of *Chelostoma* appear. They primarily examine the flowers of *C. rubra* to see if there are females in them looking for pollen or nectar, which is the case at the beginning of the reproductive season.

Shelter mimicry is a mechanism by which orchids have floral tubes in which insects can rest or sleep, hide during rain or wind, and benefit from thermoregulation, as the temperature in the floral tube can be up to 3°C higher than the ambient temperature during the morning hours (Daphni et al. 1981). This mechanism was found in the Mediterranean genus *Serapias*, whose flowers most likely mimic the entrances of beehives (Jersáková et al. 2016). Furthermore, this mechanism has been found in species of the genus *Cypripedium* (Claessens and Kleynen 2011). However, recent research has shown that *Cypripedium calceolus* also uses other mechanisms (rendezvous attraction and nest/brood site mimicry) (Claessens and Kleynen 2011).

In a mechanism popularly known as sexual deception, orchid flowers mimic female mating signals, not only by possessing attractive shape and colors, but mainly by giving off pheromone-like odors to attract males willing to mate (Jersáková et al. 2006; Pedersen and Faurholdt 2007). Flowers are pollinated by attracted male insects, which often attempt to copulate with the flower. Sexual responses range from less advanced, in which orchids deceive pollinators mainly with olfactory signals, to very complex, in which highly adapted flowers entice males to pseudocopulate (Jersáková et al. 2006; Gaskett 2011). Genera using sexual deception have evolved independently in Europe, Australia, Africa, and Central and South America (Jersáková et al. 2006; Gaskett 2011). Sexual deception has been recorded in only two South African species of the genus *Disa* – *Disa atricapilla* and *D. bivalvata* (Steiner et al. 1994). In Australia, there are several genera that use sexual deception (*Arthrochilus*, *Caleana*, *Calochilus*, *Chiloglottis*, *Cryptostylis*, *Drakaea*, *Leporella*, *Paracaleana*, *Pterostylis*, and *Spiculaea*), while in New Zealand this pollination type was found in the genera *Caladenia*, *Caleana*, *Calochilus*, *Chiloglottis*, *Cryptostylis*, and *Pterostylis* (Faegri and van der Cingel 1979; Gaskett 2011). Sexual deception has been found in *Lepanthes* species in Central America (Gaskett 2011), and it has been reported in species of five genera in South America (van der Pijl and Dodson 1966; Jersáková et al. 2006).

In Serbia, sexual deception is present in seven *Ophrys* taxa (Table 2). The low number of representatives of orchids using sexual deception in Serbia is to be expected, since this pollination system occurs in Europe only in *Ophrys* species distributed mainly in Mediterranean areas (Pedersen and Faurholdt 2007; Delforge 2016; Antonopoulos and Tsiftsis 2017) and in *Orchis galilaea* (Bino et al. 1982). Pollinators of the genus *Ophrys* can be classified into three functional groups: species with a low degree of specialization, highly specialized species, and parasitically specialized species (Vereecken et al. 2010). Sexual deception in orchids requires a high degree of specialization because insect pheromones are mostly species-specific. Moreover, many pollinators of *Ophrys* species are solitary bees belonging to the oligolectic group, i.e., a group that specializes in a particular pollen type (Dötterl and Vereecken 2010). Accordingly, specialization varies from species that attract multiple pollinator taxa to species that are pollinated by only one pollinator species.

Flowers of *Ophrys* species mimic female insects not only through the odors and volatile compounds they emit, but also through visual and tactile signals, i.e., shapes and colors (Pedersen and Faurholdt 2007). The lips of *Ophrys* flowers are often hairy and resemble the body of a female insect with folded wings, and the false nectaries resemble the eyes of an insect. In addition, recent studies suggest that reproductive success in some *Ophrys* populations is significantly increased by the presence of a strong visual signal (the conspicuous perianth) (Rakosy et al. 2012). Males are often attracted by the sex pheromone-like odor of orchids even more strongly than by the females themselves. However, males can learn to avoid areas where orchids grow, and females can increase their own attraction by moving away from the orchid population. *Ophrys* species are mostly pollinated by male solitary bees from the families Andrenidae, Apidae, Anthophoridae, Megachilidae, Colletidae, and Xylocopidae (Table 2) (Pedersen and Faurholdt 2007).

The rewarding pollination system is characteristic of 25 orchid taxa of Serbia (34.72%) from 13 genera (*Anacamptis, Coeloglossum, Epipactis, Epipogium, Goodyera, Gymnadenia, Herminium, Limodorum, Neottia, Nigritella, Platanthera, Pseudorchis*, and *Spiranthes*) (Table 2). Orchids that reward pollinators attract pollinators with nectar as well as the color, fragrance, and shape of the flower. Some orchid species produce nectar without a distinct scent (*Coeloglossum*, *Neottia*, many *Epipactis* species), while orchids that are pollinated by moths usually have strong and pleasant scents (Delforge 2006). An example is *Gymnadenia conopsea*, which is pollinated by a wide range of insects, most of which are butterflies and moths (Table 2) (Faegri and van der Cingel 1979; van der Cingel 1995; Meekers et al. 2012). Flowers attract pollinators with a scent that changes quantitatively during the diurnal cycle and is strongest at dusk, while nectar is secreted by cells in long spurs (Meekers et al. 2012). The floral scent is produced in osmophores on the lateral sepals and on the lip surface (Stpiczyńska and Matusiewicz 2001). The relative amounts of its active compounds vary from day to night. Thus, it has been found that more phenylacetaldehyde, benzaldehyde, and benzyl acetate and less benzyl benzoate are emitted at night (Huber et al. 2005).

Platanthera bifolia, whose white flowers emit an intense pleasant fragrance in the evening, is pollinated by moths, most commonly by species of the families Sphingidae and Noctuidae (Table 2) (Nilsson 1983c; Esposito et al. 2018). Nectar collects on the underside of a long spur, and the lip plays a role in visually attracting pollinators (Faegri and van der Cingel 1979). Among the rewarding orchids are some species of the genus *Epipactis*, which are most frequently pollinated by solitary and social wasps and some bees (Table 2) (van der Cingel 1995; Jakubska-Busse and Kadej 2011). Pollinators of the nectar-rewarding species *Neottia ovata* are representatives of Hymenoptera, most frequently wasps, and insects of the order Coleoptera (Table 2) (Nilsson 1981b; Kotilínek et al. 2015; Brzosko et al. 2021).

Cross-pollination, which involves a rewarding and deceptive system, occurs in the largest number of orchid taxa in Serbia (66 taxa or 91.67%), while self-pollination occurs in 11 orchid taxa (15.28%) from six genera (*Cephalanthera*, *Corallorhiza*, *Epipactis*, *Limodorum*, *Neottia*, and *Ophrys*) (Table 2). Six orchid species are exclusively self-pollinating (*Cephalanthera damasonium*, *Corallorhiza trifida*, *Epipactis leptochila* subsp. *neglecta*, *E. muelleri*, *E. pontica*, and *E. tallosii*), while five orchid taxa are both self-pollinating and cross-pollinating (*Epipactis microphylla*, *E. palustris*, *Limodorum abortivum*, *Neottia nidus-avis*, and *Ophrys apifera*). The lower number of self-pollinating orchids compared to those that are rewarding and deceptive is not surprising, as self-pollination is characteristic of a smaller number of orchid representatives. According to one study, there are 5 to 20% self-pollinating orchids (Catling 1990), while van der Pijl and Dodson (1966) found that self-pollination occurs in about 3% of orchids. The number of self-pollinating orchids increases with increasing latitude as well as in isolated geographic areas (Tremblay et al. 2005). In eastern Canada, self-pollination has been found in 17% of orchids, whereas in Europe it occurs in 27 to 50% of all orchid species (Catling 1983, 1990; Tremblay et al. 2005). It is estimated that at least 20% of the orchid species of Ecuador and Puerto Rico are autogamous (Ackerman 1983; Catling 1990).

However, the highest percentage of self-pollination (about 50%) was found in the orchids of boreal regions (Catling 1990) and on Réunion Island (Jacquemyn et al. 2005b). Orchids that are exclusively self-pollinating are rare, and most species have cross-pollination in addition to self-pollination (van der Cingel 1995).

Self-pollinating orchids in Serbia are mainly mycoheterotrophic and partially mycoheterotrophic orchids, which is consistent with the results of previous studies on orchids in Europe (Delforge 2006). Self-pollination occurs regularly in the species *Ophrys apifera*, where cross-pollination is less common. It is present for the most part in Mediterranean areas and is carried out by male bees of the genus *Eucera* Scopoli, 1770 (Fenster and Martén-Rodríguez 2007).

The pollinators of orchids of Serbia are mostly insects, while spiders (Arachnida: Araneidae) are pollinators of a small number of orchid taxa (Table 2). The largest number of orchids in Serbia (58 orchid species and subspecies) are pollinated by insects of the order Hymenoptera, followed by representatives of the orders Diptera (26), Coleoptera (22), and Lepidoptera (17), while insects from other orders (Heteroptera, Hemiptera, Orthoptera, Dermoptera, Mecoptera, and Thysanoptera) are pollinators of a smaller number of orchid taxa (Table 2).

There are several orchid taxa whose pollinators are generally poorly known. This applies to *Dactylorhiza cordigera* subsp. *cordigera*, *D. cordigera* subsp. *bosniaca*, *D. maculata* subsp. *transsilvanica*, *Gymnadenia frivaldii*, *Himantoglossum calcaratum* subsp. *calcaratum*, *H. calcaratum* subsp. *rumelicum*, and *Neotinea lactea*. Future research should therefore be devoted to detailed investigation of both the pollinators of these orchids and the pollination mechanisms themselves. Research on the pollination of orchids in Serbia is necessary because there has been almost no such research to date.

Table 2. Overview of pollination systems (R – rewarding, D – deceptive, S – self-pollination) and main pollinators of orchid taxa present in Serbia

Orchid taxon	Pollination System	Pollinators (literature)
Anacamptis coriophora subsp. *coriophora*	R	Hymenoptera (Apidae): *Apis mellifera* Linnaeus, 1758, *Eucera* spp., *Andrena* spp., *Ceratina* spp. and *Nomada* spp.; Lepidoptera: *Zygaena filipendulae* (Linnaeus, 1758); Heteroptera: *Spilostethus saxatilis* (Scopoli, 1763); and Diptera: *Tachina fera* (Linnaeus, 1761) (Dafni and Ivri 1979; van der Cingel 1995)
Anacamptis coriophora subsp. *fragrans*	R	Hymenoptera: *Apis* sp., *Andrena* sp., *Halictus* sp., *Xylocopa* sp., *Anthophora* sp. and the wasp *Trielis villosa* var. *rubra* (Vöth 1975; Dafni and Ivri 1979)
Anacamptis laxiflora	D (generalized food deception)	Hymenoptera: *Bombus* spp., *Xylocopa violacea* (Linnaeus, 1758) (http://www.giros.it/forum/viewtopic.php?f=7&t=1498)
Anacamptis morio subsp. *morio*	D (generalized food deception)	Hymenoptera: *Bombus* spp., *Andrena albopunctata* (Rossi, 1792), *Apis mellifera* Linnaeus, 1758, *Tetralonia hungarica* (Friese, 1895), *Halictus patellatus* Morawitz, 1873, *Lasioglossum xanthopus* (Kirby, 1802) and *Osmia rufa* (Linnaeus, 1758) (Johnson and Nilsson 1999; Lang 2004; Cozzolino et al. 2005)
Anacamptis morio subsp. *caucasica*	D (generalized food deception)	
Anacamptis palustris subsp. *palustris*	D (generalized food deception)	Hymenoptera: *Apis mellifera* Linnaeus, 1758 and *Bombus pascuorum* (Scopoli,1763) (van der Cingel 1995; Fritz and Nilsson 1996)
Anacamptis palustris subsp. *elegans*	D (generalized food deception)	
Anacamptis papilionacea subsp. *papilionacea*	D (rendezvous attraction)	Hymenoptera: *Eucera longicornis* (Linnaeus, 1758), *E. caspica* Morawitz, 1873, *E. nigrescens* Pérez, 1879, *Tetralonia hungarica* (Friese, 1895), *Anthophora crinipes* Smith, 1854, *A. retusa* (Linnaeus, 1758), *Chalicodoma parietina* (Geoffroy, 1785), and *Bombus humilis* Illiger, 1806 (Vogel 1972; Cozzolino et al. 2005; Scopece et al. 2009)

Table 2. (Continued)

Orchid taxon	Pollination System	Pollinators (literature)
Anacamptis pyramidalis	D (Batesian floral mimicry)	Lepidoptera: Zygaenidae, Hesperiidae, Pieridae, Nymphalidae, Sphingidae, Geometridae and Erebidae (Nazarov and Efetov 1993; Fritz and Nilsson 1996; Lang 2004; Lind et al. 2007; Molnár 2011; Claessens and Kleynen 2016c; Nahirnić and Beshkov 2018). The most important species are from the genera *Adscita* Retzius, 1783 and *Zygaena* Fabricius, 1775 – *Z. minos* ([Denis & Schiffermüller], 1775), *Z. purpuralis* (Brünnich, 1763), *Z. loti* ([Denis & Schiffermüller], 1775), *Z. lavandulae* (Esper, 1783), *Z. filipendulae* (Linnaeus, 1758) (Zygaenidae), and *Aporia crataegy* (Linnaeus, 1758) (Pieridae)
Cephalanthera damasonium	S	self-pollination (Claessens and Kleynen 2013)
Cephalanthera longifolia	D (generalized food deception)	Hymenoptera: *Lasioglossum smeathmanellum* (Kirby, 1802), *L. fulvicorne* (Kirby, 1802) and *L. laeve* (Kirby, 1802), *Dolerus* spp., and species of the genera *Halictus* Latreille, 1804 and *Andrena* Fabricius, 1775 (Dafni and Ivri 1981; Lang 2004; Claessens and Kleynen 2013)
Cephalanthera rubra	D (rendezvous attraction)	Hymenoptera: *Chelostoma rapunculi* (Lepeletier, 1841) and *C. campanularum* (Kirby, 1802), species of the genera *Heriades* Spinola, 1808 and *Osmia* Panzer, 1806 (Megachilidae), and the species of the genus *Dufourea* Lepeletier, 1841 (Halictidae) (Nilsson 1983b; Lang 2004; Sedivy et al. 2008; Vereecken et al. 2010; Claessens and Kleynen 2013)
Coeloglossum viride	R	Coleoptera: *Cantharis rustica* Fallen, 1807 (Cantharidae) and some species of the family Melyridae; Hymenoptera: species of the genus *Halictus* Latreille, 1804 (Halictidae) and *Formica exsecta* Nylander, 1846 (Formicidae); Diptera (Molnár 2011; Inda et al. 2012 and references therein; Claessens and Seifert 2017, 2018)
Corallorhiza trifida	S	self-pollination (Claessens and Kleynen 1998)
Cypripedium calceolus	D (generalized food deception, Shelter imitation, nest/brood site mimicry, rendezvous attraction)	Hymenoptera: *Andrena* spp. (Antonelli et al. 2009; Claessens and Kleynen 2013)

Orchid taxon	Pollination System	Pollinators (literature)
Dactylorhiza cordigera subsp. *cordigera*	D (generalized food deception)	Diptera, Coleoptera, Hymenoptera (Inda et al. 2012 and references therein)
Dactylorhiza cordigera subsp. *bosniaca*	D (generalized food deception)	
Dactylorhiza fuchsii subsp. *fuchsii*	D (generalized food deception)	Coleoptera: *Alosterna tabacicolor* (De Geer, 1775), *Anastrangalia reyi* (Heyden, 1889), *Pachytodes cerambyciformis* (Schrank, 1781) (Cerambycidae), *Dascillus cervinus* (Linnaeus, 1758) (Dascillidae), *Oedemera nobilis* (Scopoli, 1763), *O. lurida* (Marsham, 1802) (Oedemeridae); Diptera: some species of the family Syrphidae, *Empis tessellata* Fabricius, 1794, *Odontomyia hydroleon* (Linnaeus, 1758); Hymenoptera: *Bombus* spp., *Apis mellifera* Linnaeus, 1758 (Gutowski 1990; Lang 2004; Wróblewska et al. 2019 and references therein)
Dactylorhiza incarnata subsp. *incarnata*	D (generalized food deception)	Hymenoptera: *Apis mellifera* Linnaeus, 1758, *Bombus* spp.; Coleoptera (Lang 2004; Vallius et al. 2004; Wróblewska et al. 2019 and references therein)
Dactylorhiza maculata subsp. *maculata*	D (generalized food deception)	Hymenoptera (*Bombus* spp., some species of the family Halictidae); Diptera (Empidae, Syrphidae); Coleoptera (Nilsson 1981a; Vallius 2000; Inda et al. 2012 and references therein)
Dactylorhiza maculata subsp. *transsilvanica*	D (generalized food deception)	Hymenoptera (*Bombus* spp.) (field observation)
Dactylorhiza majalis subsp. *majalis*	D (generalized food deception)	Hymenoptera: *Bombus* spp., *Apis mellifera* Linnaeus, 1758, *Lasioglossum leucozonius* (Schrank, 1781), *Eucera longicornis* (Linnaeus, 1758), *Nomada sexfasciata* (Panzer, 1799) and *Osmia bicolor* (Schrank, 1781) (Claessens and Kleynen 2011; Ostrowiecka et al. 2019; Wróblewska et al. 2019 and references therein); Coleoptera: *Actenicerus siaelandicus* (O. F. Müller, 1764) (Claessens and Kleynen 2011)
Dactylorhiza romana subsp. *romana*	D (generalized food deception)	Hymenoptera: bumblebee queens and various other bees (Salzmann and Schiestl 2007; Inda et al. 2012 and references therein)
Dactylorhiza saccifera subsp. *saccifera*	D (generalized food deception)	Hymenoptera (*Bombus* spp., *Apis* spp.); Diptera (*Empis* spp., *Eristalis* spp.) (Inda et al. 2012 and references therein)

Table 2. (Continued)

Orchid taxon	Pollination System	Pollinators (literature)
Dactylorhiza sambucina	D (generalized food deception)	Hymenoptera: *Bombus lapidarius* (Linnaeus, 1758), *B. terrestris* (Linnaeus, 1758) and other *Bombus* species, *Apis mellifera* Linnaeus, 1758, *Andrena nigroaenea* (Kirby, 1802), *Osmia bicolor* (Schrank, 1781), *Anthophora aestivalis* (Panzer, 1801) (Nilsson 1980; van der Cingel 1995; Claessens and Kleynen 2011; Jersáková et al. 2015)
Epipactis atrorubens	R	Hymenoptera: *Bombus lucorum* (Linnaeus, 1761), *B. hortorum* (Linnaeus, 1761) (Apidae), *Dolicovespula saxonica* (Fabricius, 1793) (Vespidae), *Myrmica rubra* (Linnaeus, 1758), *M. schencki* Emeri, 1895 (Formicidae); Diptera: *Episyrphus balteatus* (De Geer, 1776) (Jakubska-Busse and Kadej 2011)
Epipactis distans	R	Hymenoptera: solitary and social wasps, some bees and ants, as well as some species of the order Coleoptera (Vöth 1999)
Epipactis helleborine subsp. *helleborine*	R	Most often solitary and social wasps and some bees (van der Cingel 1995; Jakubska-Busse and Kadej 2011). Hymenoptera: *Vespula germanica* (Fabricius, 1793), *V. vulgaris* (Linnaeus, 1758), *Dolichovespula sylvestris* (Scopoli, 1763), *Polistes gallicus* (Linnaues, 1767), *Colletes daviesanus* Smith, 1846, *Ichneumon* spp., *Myrmica ruginodis* Nylander, 1846, *Apis mellifera* Linnaeus, 1758, *Bombus bohemicus* Seidl, 1838, *Lasioglossum fratellum* (Pérez, 1903), *Bombus hortorum* (Linnaeus, 1761) and *B. hypnorum* (Linnaeus, 1758); Lepidoptera: *Amata phegea* (Linnaeus, 1758); Hemiptera: *Anaspis frontalis* (Linnaeus, 1758); Coleoptera: *Rhagonycha fulva* (Scopoli, 1763), *Coccinella septempunctata* Linnaeus, 1758, *Propylea quatuordecimpunctata* (Linnaeus, 1758), *Brassicogethes aeneus* (Fabricius, 1775), *Lagria hirta* (Linnaeus, 1758) and *Malachius bipustulatus* (Linnaeus, 1758) (Malachiidae); Diptera: *Episyrphus balteatus* (De Geer, 1776) and *Asindulum* spp.; Orthoptera; Dermaptera; Mecoptera (Jakubska-Busse and Kadej 2011; Claessens and Kleynen 2016a; Rewicz et al. 2017)
Epipactis leptochila subsp. *neglecta*	S	self-pollination (van der Cingel 1995; Claessens and Kleynen 2016a)

Orchid taxon	**Pollination System**	**Pollinators (literature)**
Epipactis microphylla	R, S	Hymenoptera: social wasps and some species of the genus *Bombus* Latreille, 1802 and self-pollination (Bonatti et al. 2006; Claessens and Kleynen 2014)
Epipactis muelleri subsp. *muelleri*	S	self-pollination (van der Cingel 1995; Claessens and Kleynen 2014, 2016a)
Epipactis palustris	R, S	Self-pollination; Hymenoptera, Diptera, Coleoptera and Lepidoptera, as well as spiders (Arachnida: Araneidae) (Jacquemyn et al. 2014). Hymenoptera: *Apis mellifera* Linnaeus, 1758, *Bombus lapidarius* (Linnaeus, 1758), *B. lucorum* (Linnaeus, 1761), *B. pascuorum* (Scopoli, 1763), *Formica rufibarbis* Fabricius, 1793, *F. fusca* Linnaeus, 1758; Diptera: *Syritta pipiens* (Linnaeus, 1758), *Episyrphus* sp., *Herina frondescentiae (*Linnaeus, 1758), *Empis* sp. (Nilsson 1978; Jakubska-Busse and Kadej 2011; Jacquemyn et al. 2014)
Epipactis pontica	S	self-pollination (Delforge 2006)
Epipactis purpurata	R	Hymenoptera: *Vespula austriaca* (Panzer, 1799), *V. vulgaris* (Linnaeus, 1758) (Vespidae), Chrysididae; Diptera: *Episyrphus balteatus* (De Geer, 1776) (Syrphidae); Coleoptera: *Coccinella septempunctata* Linnaeus, 1758 (Jakubska-Busse and Kadej 2011)
Epipactis tallosii subsp. *tallosii*	S	self-pollination (Delforge 2006)
Epipogium aphyllum	R	Hymenoptera: *Bombus lucorum* (Linnaeus, 1761), *B. pascuorum* (Scopoli, 1763), *B. soroeensis proteus* Gerstäcker, 1869 (Taylor and Roberts 2011; Molnár 2011)
Goodyera repens	R	Hymenoptera: *Bombus lapidarius* (Linnaeus, 1758), *B. terrestris* (Linnaeus, 1758), *B. pascuorum* (Scopoli, 1763) (Claessens and Kleynen 2016b)
Gymnadenia conopsea	R	Lepidoptera: Hesperiidae, Pieridae, Lycaenidae and Nymphalidae; Zygaenidae, Sphingidae, Geometridae and Noctuidae. The following species stand out: *Autographa gamma* (Linnaeus, 1758), *A. bractea* ([Denis & Schiffermüller], 1775), *Mythimna conigera* ([Denis & Schiffermüller], 1775) (Noctuidae); *Macroglossum stellatarum* (Linnaeus, 1758), *Deilephila porcellus* (Linnaeus, 1758) (Sphingidae); *Thymelicus lineola* (Ochsenheimer, 1808), *Ochlodes sylvanus* (Esper, 1777) (Hesperiidae); *Vanessa cardui*

Table 2. (Continued)

Orchid taxon	Pollination System	Pollinators (literature)
		(Linnaeus, 1758) (Nymphalidae); and *Zygaena* spp. (Zygaenidae) (van der Cingel 1995; Nazarov and Buchsbaum 2004; Huber et al. 2005; Meekers et al. 2012; Hedrén et al. 2018a); Hymenoptera: *Bombus terrestris* (Linnaeus, 1758), *B. pascuorum* (Scopoli, 1763); Coleoptera: *Paracorymbia fulva* De Geer, 1775; Diptera: *Empis livida* Linnaeus, 1758, *E. tessellata* Fabricius, 1794; Hemiptera: *Spilostethus saxatilis* (Scopoli, 1763) (Claessens and Kleynen 2011; Meekers et al. 2012)
Gymnadenia frivaldii	R	Lepidoptera (Bateman et al. 2006)
Gymnadenia odoratissima	R	Lepidoptera: *Polyommatus coridon* (Poda, 1761) (Lycaenidae); *Eana osseana* (Scopoli, 1763) (Tortricidae); *Platyptilia gonodactyla* (Denis & Schiffermüller, 1775) (Pterophoridae), some species of the family Pyralidae and Geometridae (Huber et al. 2005), *Adscita statices* (Linnaeus, 1758) (Vöth 2000), *Zygaena minos* ([Denis & Schiffermüller), 1775] (Nazarov and Buchsbaum 2004) (Zygaenidae), *Eudonia* spp. (Sun et al. 2014). Some species of the orders Diptera and Coleoptera (Sun et al. 2014)
Herminium monorchis	R	Hymenoptera: females of parasitic wasps of the genus *Tetrastichus* Haliday, 1844 (Nilsson 1979); Diptera, Coleoptera (Inda et al. 2012 and references therein)
Himantoglossum calcaratum subsp. *calcaratum*	D (generalized food deception)	Hymenoptera: female bees (Inda et al. 2012 and references therein)
Himantoglossum calcaratum subsp. *rumelicum*	D (generalized food deception)	Hymenoptera: female bees (Inda et al. 2012 and references therein)
Limodorum abortivum	R, S	Self-pollination; Hymenoptera: *Anthophora* Latreille, 1803, *Rhodanthidium septemdentatum* (Latreille, 1809), *Lasioglossum* spp., *Bombus terrestris* (Linnaeus, 1758) (Claessens and Kleynen 2011, 2014)
Neotinea lactea	D (generalized food deception)	Diptera (Inda et al. 2012 and references therein)
Neotinea tridentata subsp. *tridentata*	D (generalized food deception)	Hymenoptera: *Apis mellifera* (Linnaeus, 1758), *Halictus patellatus* Morawitz 1873, *Osmia rufa* (Linnaeus 1758), *O. niveata* Fabricius, 1804 (Cozzolino et al. 2005)

Orchid taxon	**Pollination System**	**Pollinators (literature)**
Neotinea ustulata	D (generalized food deception)	Diptera: *Tachina magnicornis* (Zetterstedt, 1844) (Tachinidae); Lepidoptera: some species of the family Noctuidae and Erebidae; Coleoptera: *Leptura livida* (Fabricius, 1776) (Cerambycidae) (Vöth 1984; van der Cingel 1995; Lang 2004; Tali et al. 2004)
Neottia cordata	R	Diptera: Mycetophilidae, Sciaridae and Tipulidae; Hymenoptera: Braconidae, Ichneumonidae (Ackerman and Mesler 1979; Kotilínek et al. 2018)
Neottia nidus-avis	R, S	Self-pollination; Diptera; Thysanoptera (Burns-Balogh et al. 1987; Lang 2004)
Ophrys apifera	S, D (sexual deception)	Self-pollination; Hymenoptera: male bees of the genus *Eucera* Scopoli, 1770, mostly *E. punctulata* Alfken, 1942 (Fenster and Martén-Rodríguez 2007; Gaskett 2011 and references therein)
Ophrys fuciflora	D (sexual deception)	Hymenoptera: *Eucera nigrescens* Pérez, 1879, *Eucera longicornis* (Linnaeus,1758); Diptera: *Microdon analis* (Macquart, 1842), *M. mutabilis* (Linnaeus, 1758); Coleoptera: *Phyllopertha horticola* (Linnaeus, 1758) (Delforge 2006; Gaskett 2011 and references therein)
Ophrys insectifera subsp. *insectifera*	D (sexual deception)	Hymenoptera: male wasps of the genus *Argogorytes* Ashmead, 1899 – *A. mystaceus* (Linnaeus, 1761) and *A. fargeii* (Shuckard, 1837) (van der Cingel 1995; Pedersen and Faurholdt 2007; Vereecken et al. 2010; Gervasi et al. 2017); Coleoptera: *Eusphalerum minutum* Fabricius, 1792 (Gaskett 2011 and references therein)
Ophrys sicula	D (sexual deception)	Hymenoptera: *Andrena hesperia* Smith, 1853, *A. vulpecula* Kriechbaumer, 1873, *A. merula* Warncke, 1969, *A. taraxaci* Giraud, 1861, *A. bicolor* Fabricius, 1775 (Delforge 2006; Gaskett 2011 and references therein)
Ophrys scolopax subsp. *cornuta*	D (sexual deception)	Hymenoptera: male bees (Apidae) – *Eucera longicornis* (Linnaeus, 1758), *E. interrupta* Bär, 1850, *E. nigrescens* Pérez, 1879 (Gaskett 2011 and references therein; Molnár 2011)
Ophrys sphegodes subsp. *sphegodes*	D (sexual deception)	Hymenoptera: male bees (Apidae) – *Andrena nigroaenea* (Kirby, 1802), *A. limata* Smith, 1853, *A. cineraria* (Linnaeus, 1758), *A. barbilaris* (Kirby 1802) (Schiestl et al. 2000; Gaskett 2011 and references therein; Molnár 2011)
Ophrys sphegodes subsp. *taurica*	D (sexual deception)	

Table 2. (Continued)

Orchid taxon	Pollination System	Pollinators (literature)
Orchis mascula subsp. *mascula*	D (generalized food deception)	Hymenoptera: *Bombus* spp., *Bombylius major* Linnaeus, 1758, *Psithyrus* spp., *Osmia bicolor* (Schrank, 1781), *Argogorytes mystaceus* (Linnaeus, 1761), *Tetralonia hungarica* (Friese, 1896), *Eucera longicornis* (Linnaeus,1758), *Halictus sexcinctus* Fabricius, 1775, *Nomada marshamella* (Kirby, 1802) and *Andrena* spp. (Nilsson 1983a; Johnson and Nilsson 1999; Cozzolino et al. 2005; Jacquemyn et al. 2009); Lepidoptera: *Zygaena transalpina* (Esper, 1780) and *Minucia lunaris* (Denis & Schiffermüller, 1775); Diptera: *Empis tessellata* Fabricius, 1794, *Scathophaga stercoraria* (Linnaeus, 1758) and some species of the family Anthomyiidae (Nilsson 1983a; Jacquemyn et al. 2009)
Orchis mascula subsp. *speciosa*	D (generalized food deception)	
Orchis militaris subsp. *militaris*	D (generalized food deception)	Hymenoptera: *Andrena enslinella* Stoeckhert, 1924, *A. hattorfiana* (Fabricius, 1775), *Apis mellifera* Linnaeus, 1758, *Nomada striata* Fabricius,1793, *Bombus* spp., *Halictus eurygnathus* Blüthgen,1931, *H. simplex* Blüthgen, 1923, *Osmia aurulenta* (Panzer, 1799), *O. bicolor* (Schrank, 1781) (Vöth 1987; Lang 2004; Claessens and Kleynen 2011; Molnár 2011; Bateman and Rudall 2014). Diptera: Syrphidae, Muscidae; Lepidoptera: *Pieris* spp. (Farrell 1985; Lang 2004; Claessens and Kleynen 2011)
Orchis pallens	D (Batesian floral mimicry)	Hymenoptera: *Bombus pascuorum* (Scopoli, 1763), *B. pratorum* (Linnaeus, 1761) and *B. terrestris* (Linnaeus, 1758) (Vöth 1982; Dafni 1987)
Orchis provincialis	D (generalized food deception, Batesian floral mimicry)	Hymenoptera: *Andrena lathyri* Alfken, 1899 (Svolynskiy et al. 2014)
Orchis purpurea subsp. *purpurea*	D (generalized food deception)	Hymenoptera: *Ancistrocerus parietum* (Linnaeus, 1758), *Helictus* spp., *Bombus terrestris* (Linnaeus, 1758), *Apis mellifera* Linnaeus, 1758; Lepidoptera: *Iphiclides podalirius* (Linnaeus, 1758); Coleoptera: *Tropinota hirta* (Poda, 1761) (Lang 2004; Fantinato et al. 2017; Molnár 2011)
Orchis simia subsp. *simia*	D (generalized food deception)	Hymenoptera: some species of the genera *Andrena*, *Apis*, *Halictus* and *Lasioglossum*; Diptera: *Bombylius major* Linnaeus, 1758 and *Scathophaga* species; Coleoptera: *Tropinota hirta* Poda, 1761 (Lang 2004; Molnár 2011; Fantinato et al. 2017)

Orchid taxon	**Pollination System**	**Pollinators (literature)**
Orchis spitzelii subsp. *spitzelii*	D (generalized food deception)	Hymenoptera: *Bombus* spp., *Psithyrus* spp. (Apidae) (Fritz 1990)
Platanthera bifolia	R	Lepidoptera: Sphingidae and Noctuidae (Nilsson 1983c; Esposito et al. 2018)
Platanthera chlorantha subsp. *chlorantha*	R	Lepidoptera: Sphingidae, Noctuidae and Geometridae (Nilsson 1983c; Hausmann 2004; Esposito et al. 2018)
Pseudorchis albida subsp. *albida*	R	Lepidoptera: *Crambus* Fabricius, 1798 (Crambidae) and *Hellinsia* Tutt, 1905 (Pterophoridae); Diptera: *Empis tessellata* Fabricius, 1794 and *E. bistortae* Meigen, 1822 (Empididae); Hymenoptera: some solitary bees (van der Pijl and Dodson 1966; Jersáková et al. 2011)
Spiranthes aestivalis	R	Hymenoptera: honeybees and bumblebees (Schatz et al. 2004)
Spiranthes spiralis	R	Hymenoptera: *Bombus terrestris* Linnaeus, 1758, *B. lapidarius* Linnaeus, 1758, *B. sylvarum* (Linnaeus, 1761), *Halictus simplex* Blüthgen, 1923 and *Apis mellifera* Linnaeus, 1758 (Apidae) (Jacquemyn and Hutchings 2010)
Traunsteinera globosa	D (Batesian floral mimicry)	Diptera: Empididae and Syrphidae; Hymenoptera: bees; Lepidoptera: Nymphalidae and Noctuidae; Heteroptera: Lygaeidae; Coleoptera: Cerambycidae, Oedemeridae (Jersáková et al. 2016)

PHYTOGEOGRAPHICAL ANALYSIS

Chorological analysis of the orchid flora in Serbia revealed the presence of nine basic chorological groups (Figure 19). After classification into chorological groups and subgroups, its belonging to a category of floristic elements was determined for each orchid taxon based on geography of the taxon´s range. The chorological types for phytogeographical analysis were determined according to the principles defined by Meusel et al. (1965, 1978), Meusel and Jäger (1992), and Stevanović (1992). The presence of orchids from different chorological groups was expected, in view of the fact that Serbia is located in an area influenced by different floristic-vegetation regions (Stevanović 1995).

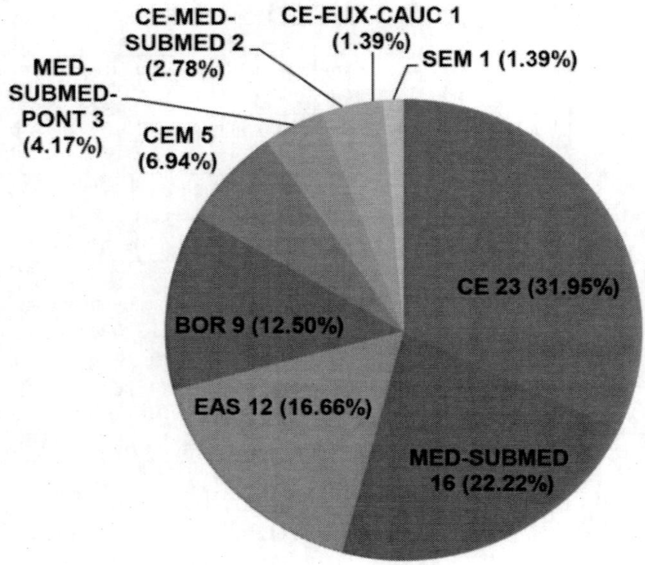

Figure 19. Spectrum of basic chorological groups of the orchid flora in Serbia: CE – Central European; MED-SUBMED – Mediterranean-Submediterranean; EAS – Eurasian, BOR – Boreal; CEM – Central European mountainous; MED-SUBMED-PONT – Mediterranean-Submediterranean-Pontic; CE-MED-SUBMED – Central European-Mediterranean-Submediterranean; CE-EUX-CAUC – Central European-Euxine-Caucasian; SEM – South European mountainous.

This can be explained by geomorphological, geological, and historical features, significant altitudinal differences, and different vegetation types in the studied area (Tomović et al. 2014; Jakovljević et al. 2020). The corresponding chorological groups, chorological subgroups, and floristic elements of Serbian orchid taxa are presented in Table 3.

Chorological analysis of the orchid flora in Serbia (Figure 19) shows a pronounced dominance of orchids belonging to the Central European chorological group (23 taxa; 31.95%) from 11 genera (*Anacamptis*, *Cephalanthera*, *Dactylorhiza*, *Epipactis*, *Gymnadenia*, *Neotinea*, *Neottia*, *Ophrys*, *Orchis*, *Platanthera*, and *Spiranthes*) (Table 3). The high number of Central European orchid representatives is understandable in view of the fact that a significant part of Serbia has a temperate-continental climate and a large presence of oak, beech, hornbeam, and beech-fir forests, as well as mesophilous and hygrophilous herbaceous ecosystems, where the majority of orchids of this chorological group grow (Djordjević et al. 2016b, 2020a, 2020b). Among representatives of the group, the genus *Epipactis* stands out, with a total of seven taxa (Table 3). During the Great Ice Age, *Epipactis* species had a limited distribution in Southern Europe, i.e., on the Iberian, Apennine, and Balkan Peninsulas, as well as in the Caucasus (Delforge 2006). Deforge assumes that a process of evolutionary radiation within this genus occurred in the area of Central Europe during the post-glacial period, when slow immigration of beech forests from Southern Europe took place. It is worth highlighting the presence of *Dactylorhiza maculata* subsp. *transsilvanica*, which is a Carpathian-Balkan subendemic taxon. This taxon has in Serbia the southern limit of its distribution in Europe.

In second place is the Mediterranean-Submediterranean group, with 16 taxa (22.22%) from seven genera (*Anacamptis*, *Dactylorhiza*, *Himantoglossum*, *Limodorum*, *Neotinea*, *Ophrys*, and *Orchis*) (Table 3). In addition, three representatives of the Mediterranean-Submediterranean-Pontic group and two orchids of the Central European-Mediterranean-Submediterranean chorological group are present in Serbia (Table 3). The significant number of orchids from the Mediterranean-Submediterranean group is understandable, since the southern part of Serbia is relatively close to the Adriatic-Ionian province within the European-Submediterranean floristic-vegetation subregion (Stevanović 1995). In addition, the studied region has areas of lower elevations dominated by thermophilous vegetation types, such as forest communities of the class *Quercetea pubescentis*, order *Quercetalia pubescenti-petraeae*, and (in particular) the alliances *Fraxino orni-Ostryion*, *Quercion petraeo-cerridis*, and *Quercion confertae* (Djordjević et al. 2020a, 2020b). The significant number of these orchids can also be explained by the large presence of thermophilous grassland communities of the class *Festuco-Brometea*, especially within the orders *Brachypodietalia pinnati* and *Festucetalia valesiacae* (Djordjević et al. 2016b, 2020b). It is important to note that many representatives of Mediterranean-Submediterranean orchids occur in limestone gorges and canyons of Serbia, which can be explained by the fact that these areas are protected from extreme climatic influences and consequently are characterized by conditions of temperature, humidity, and light favorable for the development and survival of orchids (Djordjević et al. 2020b). Moreover, limestones are generally suitable for most orchids (Djordjević and Tsiftsis 2019).

Among the representatives of this chorological group, those species that have their northern limit of distribution in this part of Europe stand out. The first of these is *Anacamptis papilionacea*, a species that occurs mainly in "steppe-like" communities of the alliance *Chrysopogono-Danthonion calycinae* on carbonates and ultramafics (Djordjević et al. 2016b, 2020b). The second is *Dactylorhiza saccifera*, which, as a typical species of Southern Europe, is a very common, ecologically plastic species in Serbia, inhabiting both forest and grassland ecosystems (Djordjević et al. 2020b). The localities of *Himantoglossum calcaratum* subsp. *calcaratum* in northwestern Serbia represent the northernmost points of its distribution on the Balkan Peninsula. This taxon is an endemic taxon of the Balkan Peninsula, distributed in eastern Bosnia and Herzegovina, Montenegro, Serbia, northern Albania, and North Macedonia (Delforge 2006; Fekete et al. 2017; Djordjević et al. 2018). In Serbia, it is distributed mainly in the western part of the country, i.e., in northwestern, western, and southwestern Serbia, then in Šumadija, central Serbia, and Kosovo and Metohija (Djordjević et al. 2016b, 2018, 2020a). The abundance and significant presence of this taxon are particularly high in the vicinity of Prijepolje and Priboj (Djordjević 2018). On the other hand, *Himantoglossum calcaratum* subsp. *rumelicum*, which belongs to the Mediterranean-Submediterranean-Pontic chorological group, has a wider distribution, more precisely extending from southern Slovakia to northwestern Turkey (WCSP 2020). In Serbia, this taxon is distributed in Vojvodina (the northern part of Serbia), as well as in northeastern, eastern, southeastern, and southern Serbia and Šumadija (Djordjević et al. 2018). However, the boundaries of the ranges of these two *Himantoglossum* taxa in Serbia are not clear and require further research.

Representatives of the typical Mediterranean genus *Serapias* are absent in Serbia, as are many species of the genera *Anacamptis*, *Himantoglossum*, *Ophrys*, and *Orchis* (*Anacamptis sancta*, *Himantoglossum robertianum*, *Orchis anthropophora*, *O. quadripunctata*, *O. pauciflora*, *Orchis punctulata*, etc.) that occur in the Mediterranean and sub-Mediterranean areas of Southern Europe, including Balkan countries such as Albania, Bosnia and Herzegovina, Croatia, Montenegro, Greece, and Turkey (Delforge 2006; Tsiftsis and Antonopoulos 2017; Antonopoulos and Tsiftsis 2017; Tsiftsis and Djordjević 2018, 2020).

The Eurasian chorological group includes 12 taxa (16.66%) from nine genera (*Anacamptis*, *Cephalanthera*, *Cypripedium*, *Epipactis*, *Gymnadenia*, *Herminum*, *Neottia*, *Orchis*, and *Platanthera*) (Table 3). Many orchids in this chorological group have great ecological plasticity and adaptability, which allows them to survive in different habitat types. Moreover, a great number of them are known to inhabit North Africa in addition to Europe and Asia. The most widespread and least specialized orchids of this chorological group include *Cephalanthera longifolia*, *Epipactis helleborine*, *Gymnadenia conopsea*, *Neottia ovata*, and *Platanthera bifolia* (Buttler 1991; Delforge 2006; Djordjević et al. 2020a, 2020b).

The boreal chorological group has a significant presence – nine orchid taxa (12.50%) from six genera (*Corallorhiza*, *Coeloglossum*, *Dactylorhiza, Epipogium*, *Goodyera*, and *Neottia*) (Table 3). This is understandable in view of the presence of a humid mountain climate of the alpine type in Serbia (Stevanović and Stevanović 1995). The boreal floristic-vegetation region corresponds to the Holarctic taiga from the point of view of the ecosystem in question, while in Serbia it is represented by the Illyrian province within the Balkan-boreal subregion (Stevanović 1995). The presence of boreal orchids in Serbia is attributable to the extensive distribution of coniferous forests of the classes *Vaccinio-Piceetea* and *Erico-Pinetea*, as well as fen communities of the class *Scheuchzerio palustris-Caricetea fuscae*, which are important habitats where representatives of this chorological group of orchids have their greatest abundances (Djordjević et al. 2016b, 2020a, 2020b; Tsiftsis et al. 2019a). In addition, recent studies have shown that important habitats where boreal orchids grow are hygrophilous and mesophilous grassland communities from the class *Molinio-Arrhenatheretea*, as well as communities from the classes *Nardetea strictae* and *Juncetea trifidi* (Djordjević et al. 2016b, 2020b). Many boreal orchids in Serbia are also recorded in beech and beech-fir forests of the class *Carpino-Fagetea sylvaticae*, *Picea omorika* forests (*Erico carneae-Piceion omorikae*), and pine forests (*Erico-Fraxinion orni*) on ultramafics and cold mountain plateaus (Djordjević et al. 2020a, 2020b). Recent studies have shown that in western Serbia *Corallorhiza trifida* is an indicator species of *Picea abies* forests, while *Goodyera repens* is an indicator species of *Picea omorika* forests (Djordjević et al. 2020a). The significant occurrence of boreal orchids in Serbia can also be explained by the widespread occurrence of siliceous geological substrates. Recent research in the Central Balkans has indicated that silicate substrates in a broader sense favor many Arctic-Alpine species (Stevanović et al. 2009) and boreal species (Vukojičić et al. 2014), including numerous boreal orchids (Djordjević and Tsiftsis 2019). This can be explained by the fact that most silicate substrates, especially acidic and intermediate igneous rocks, have a high water-storage capacity, and by the fact that silicate substrates, which are suitable for many representatives of boreal orchids, occupy large areas at higher elevations. Three taxa of *Dactylorhiza* should be mentioned (Table 3), the localities of two of which (*Dactylorhiza fuchsii* and *Dactylorhiza maculata* subsp. *maculata*) represent the southernmost limit of their distribution in the Balkans (Djordjević et al. 2014; Djordjević 2018).

The Central European mountainous group is represented by five taxa (6.94%) from four genera (*Dactylorhiza*, *Gymnadenia*, *Nigritella*, and *Traunsteinera*) (Table 3). A lower number of orchid representatives from this chorological group is to be expected, as there are few areas above 1800 m in Serbia. Among the group´s representatives, the Balkan endemic *Dactylorhiza cordigera* subsp. *bosniaca* should be highlighted (Tomović et al. 2014), together with two subendemics of the Carpathians and the Balkans (*Gymnadenia frivaldii* and *Dactylorhiza cordigera* subsp. *cordigera*) (Delforge 2006; Bateman et al. 2006). Moreover, it is important to emphasize that the localities of *Traunsteinera globosa* in Serbia represent its southernmost distribution limit in this part of Europe.

Table 3. Overview of orchids of Serbia with indication of chorological groups, subgroups, and floristic elements

Orchid taxon	Chorological groups	Chorological subgroups	Floristic elements
Anacamptis coriophora subsp. *coriophora*	CE	CE(W-E)-MED-SUBMED-OR-PONT	CE(W-E)(atl-CE-sarm)-pont-med-submed(W-E)-or-anatol
Anacamptis coriophora subsp. *fragrans*	MED-SUBMED	MED-SUBMED(W-E)-OR	Atl(S)-med-submed(W-E)-or(kurd-iran(NW))
Anacamptis laxiflora	MED-SUBMED	MED-SUBMED(W-E)	atl(S)-med-submed(W-E)
Anacamptis morio subsp. *morio*	CE	CE-(med-submed-pont)	atl-med-submed(C-E)-CE(W-E)-pont(W)
Anacamptis morio subsp. *caucasica*	MED-SUBMED-PONT	MED-SUBMED-OR-PONT-TURAN	pont(W-E)-illyr-balk-ion-pelop-hellen-aeg-cret-anatol-or-hyrc
Anacamptis palustris subsp. *palustris*	EAS	EAS(W-C)(temp-submerid)	atl(S)-CE(W-E)-eux-cauc-pont-iran-turan-aral-casp
Anacamptis palustris subsp. *elegans*	EAS	EAS(W-C)-temp-submerid	CE(hyrc-polon-sarm)-illyr-balk-pont-sib-med-submed(E)-or-turan(aral-casp)
Anacamptis papilionacea subsp. *papilionacea*	MED-SUBMED	MED-SUBMED(W-E)	afr(NW)-Atl(S)-med-submed(W-E)-dac(S)-anatol
Anacamptis pyramidalis	MED-SUBMED	MED-SUBMED-PONT-TURAN-(CE)	atl-med-submed(W-E)-CE(W-SE)-pont(W-C)-crim-eux-cauc-iran(N)
Cephalanthera damasonium	CE	CE-CAUC-(med-submed)	atl-med-submed(W-E)-CE(W-E)-pont(W)-cauc-iran(N)
Cephalanthera longifolia	EAS	EAS(W-C)	atl-med-submed(W-E)-CE(W-E)-cauc-him(W)
Cephalanthera rubra	CE	CE-CAUC-(med-submed)	CE(W-E)-med-submed(W-E)-eux-cauc
Coeloglossum viride	BOR	CIRCBOR	NAm-EAS (bor-temp)
Corallorhiza trifida	BOR	CIRCBOR	NAm-EAS (bor-temp)

Table 3. (Continued)

Orchid taxon	Chorological groups	Chorological subgroups	Floristic elements
Cypripedium calceolus	EAS	EAS(W-E)(bor-temp)	CE(subatl-herc.balt-sarm-alp-illyr-carp)-borscand-borrus-borsib(W)-sib(C)-daur-manch
Dactylorhiza cordigera subsp. *cordigera*	CEM	CEM(C-E)	carp-balk(moes)
Dactylorhiza cordigera subsp. *bosniaca*	CEM	CEM(C-E)	dinar(C-E)-balk(sc-pind(N)-moes(W-C))(salp-alp)
Dactylorhiza fuchsii subsp. *fuchsii*	BOR	EUSIB(W-C)(arc-submerid)	atl-CE(W-E)-eubor-cauc-sib(S)
Dactylorhiza incarnata subsp. *incarnata*	BOR	EUSIB(W-C)(arc-bor-submerid)	atl-CE(W-E)-eux-cauc-iran(N)-scand-sib(W-C)-him(W)
Dactylorhiza maculata subsp. *maculata*	BOR	EUSIB(W-E)(bor-submerid)	atl-CE(W-E)-sib(W-E)
Dactylorhiza maculata subsp. *transsilvanica*	CE	CE(SE)	carp-dinar-balk(moes(W))
Dactylorhiza majalis subsp. *majalis*	CE	CE(W-E)	atl-CE(W-E)
Dactylorhiza romana subsp. *romana*	MED-SUBMED	MED-SUBMED(W-E)	afr(NW)-iber-cors-sard-sic-apen-adr-illyr-ion-aeg-cret-balk-crim-anatol-leb-syr
Dactylorhiza saccifera subsp. *saccifera*	MED-SUBMED	MED-SUBMED(W-E)	iber-cors-sic-apen-illyr-balk-dac(S)-anatol
Dactylorhiza sambucina	CE	CE(W-E)-CAUC	CE(W-E)-cauc-iran(N)-submed(mont-subalp)
Epipactis atrorubens	EAS	EAS(W-C)(bor-temp)	atl-CE(W-E)-eubor-cauc-sib(W-C)
Epipactis distans	CE	CE(W-C)(temp)	iber(N)-alp-apen(N)-illyr-bohem-balt
Epipactis helleborine subsp. *helleborine*	EAS	EAS(W-E)(temp-submerid)	atl-med-submed(W-E)-CE(W-E)-cauc-pont(W)-sib(W-E)-alt-him(W-C)
Epipactis leptochila subsp. *neglecta*	CE	CE(C-SE)	alp(W-E)-dinar(W)-herc-balk(moes)
Epipactis microphylla	CE	CE-CAUC-(med-submed)	CE(W-C)-cauc-iran(N)-med-submed(W-E)

Orchid taxon	Chorological groups	Chorological subgroups	Floristic elements
Epipactis muelleri subsp. *muelleri*	CE	CE(W-C)-med-submed(W-C)	CE(W-C)-apen-illyr(W-E)-herc-iber(NE)-sard-apen-adr(N)
Epipactis palustris	EAS	EAS(W-C)(subbor-temp)	atl-CE(W-E)-scand-cauc-alta-sib(W-C)
Epipactis pontica	CE	CE-EUX	herc-illyr(W)-balk(moes)-eux
Epipactis purpurata	CE	CE(W-E)-CAUC	Atl-CE(W-E)-cauc
Epipactis tallosii subsp. *tallosii*	CE	CE(SC)	matr-pann
Epipogium aphyllum	BOR	EUSIB(W-E)(bor-temp)	eubor-sib(W-E)-kam-jap-EASM (pyr-alp-apen-dinar-balk-carp-cauc-iran(N)-him(W-C))
Goodyera repens	BOR	CIRCBOR	NAm(bor)-sib(W-E)-kam-jap-EASM
Gymnadenia conopsea	EAS	EAS(W-E)(bor-temp)	atl-CE-arc-borscand-sib(W-E)(bor-temp)-jap(N)
Gymnadenia frivaldii	CEM	CEM(C-E)	dinar(CE)-balk(sc-pind(N)-moes(W))-carp
Gymnadenia odoratissima	CE	CE(W-E)	subatl(S)-alp-CE-illyr-balk-balt
Herminium monorchis	EAS	EAS(W-E)(temp)	subatl-CE(W-E)-cauc-eusib(W-E)(subbor-temp)-him-chin
Himantoglossum calcaratum subsp. *calcaratum*	MED-SUBMED	SUBMED-SUBCONT(E)	illyr-balk(sc-pind(N))
Himantoglossum calcaratum subsp. *rumelicum*	MED-SUBMED-PONT	MED-SUBMED(E)-PANN	pann-balk(moes(W-E)-maced-thrac-hellen-aeg-pelop)-anatol(N)
Limodorum abortivum	MED-SUBMED	MED-SUBMED (W-E)	atl(S)-med-submed(W-E)-or-pont(W)
Neotinea lactea	MED-SUBMED	MED-SUBMED (W-E)	afr(NW-NC)-gal-cors-sard-sic-apen-illyr-ion-aeg-pelop-cret-illyr-balk-anatol
Neotinea tridentata subsp. *tridentata*	MED-SUBMED	MED-SUBMED(med-submed-CE-cauc-pont)	med-submed(W-E)-CE(S)-cauc-eux-pont(w)-crim
Neotinea ustulata	CE	CE(W-E)	atl-CE(W-E)-pont(W)-sib(SW)
Neottia cordata	BOR	CIRCBOR	NAm-EAS(bor-temp)
Neottia nidus-avis	CE	CE-CAUC	atl-CE(W-E)-cauc
Neottia ovata	EAS	EAS(W-C)(bor-submerid)	atl-CE(W-E)-cauc-scand(Bor)-sib(SW-SC)-alta-him(W)
Ophrys apifera	MED-SUBMED	MED-SUBMED-(CE-eux-crim)	atl-med-submed(W-E)-CE(W)-eux-crim-iran(N)

Table 3. (Continued)

Orchid taxon	Chorological groups	Chorological subgroups	Floristic elements
Ophrys fuciflora	CE-MED-SUBMED	CE(W-SC)-MED-SUBMED(C-E)	Atl-subatl-herc(SC)(-pann-cors-sard-sic-afr(NC)-apen-adr-illyr-aeg-cret-anatol(W-S)-cyp-leb
Ophrys insectifera subsp. *insectifera*	CE	CE(W-C)	atl-CE(W-C)-submed(W-C)
Ophrys sicula	MED-SUBMED	MED-SUBMED(W-E)	afr(NW-NC)-iber-cors-sard-sic-apen-adriat-illyr-ion-aeg-maced-thrac-cret-anatol
Ophrys scolopax subsp. *cornuta*	MED-SUBMED	MED-SUBMED(W-E)	W-E med-submed-or-turan-pont-cauc
Ophrys sphegodes subsp. *sphegodes*	MED-SUBMED	MED-SUBMED(W-E)-EUX-CAUC	atl-med-submed(W-E)-crim-eux-cauc
Ophrys sphegodes subsp. *taurica*	MED-SUBMED-PONT	MED-SUBMED(E)-PONT(W-C)	adr-illyr-balk(sc-pind(N-S)-moes(SW-E)-ion-aeg-pelop)-cret-anatol-pann-danub-crim
Orchis mascula subsp. *mascula*	CE	CE-EUX-CAUC-MED-SUBMED	atl-med-submed(W-E)-CE(W-E)-eux-crim-cauc-iran(N)
Orchis mascula subsp. *speciosa*	CE-EUX-CAUC	CE(C-E-SE)-EUX-CAUC-HYRC	polon-sarm-alp(E)-illyr-balk(sc-pind(N)-moes(W-E))-carp(N-S)-eux-cauc-hyrc
Orchis militaris subsp. *militaris*	EAS	EAS(W-C) (temp-merid)	CE(W-SE)-cauc-pont(W)-alta-sib(C)
Orchis pallens	CE	CE-CAUC	CE(W-E)-cauc-pont(W)
Orchis provincialis	MED-SUBMED	MED-SUBMED(W-E)-CRIM-CAUC	lusit(N-S)-galec-cant-iber(N)-catal-provenc-cors-sic-pad-apen-adr-illyr(W)-alb-hellen(W)-aeg-cret-maced-thrac-anatol(W-N)-eux-crim-cauc
Orchis purpurea subsp. *purpurea*	CE	CE-CAUC	atl(S)-CE(W-SE)-submed(C-NE)-pont(W)-crim-cauc
Orchis simia subsp. *simia*	MED-SUBMED	MED-SUBMED(W-E)-CRIM-CAUC	atl-med-submed(C-E)-crim-cauc
Orchis spitzelii subsp. *spitzelii*	SEM	SEM(W-E)	pyr-alp-apen-dinar-balk-carp-anatol-balt
Platanthera bifolia	EAS	EAS (W–C) (temp-submerid)	atl-med-submed (W-E)-CE-cauc-sib (SW-C)
Platanthera chlorantha subsp. *chlorantha*	CE	CE-CAUC	atl-aed-submed (W-E)-CE-cauc

Orchid taxon	Chorological groups	Chorological subgroups	Floristic elements
Pseudorchis albida subsp. *albida*	BOR	EAS(W)	ice-boratl-borrus-atl-herc-CEP(pyr-alp-dinar-balk(sc-pind(N)-moes(W-SW-C))-tatr-carp(N-S)(arc-temp/salp-alp)
Spiranthes aestivalis	CE-MED-SUBMED	CE(W-C-SE)-MED-SUBMED(W-C)	atl-subatl-herc-polon(S)-afr(NW)-cors-sard-apen(N-C)-alp-illyr-balk(sc-pind(S))-carp
Spiranthes spiralis	CE	CE-CAUC	atl-CE(W-C)-submed(W-E)-cauc
Traunsteinera globosa	CEM	CEM(W-E)-ANATOL	pyr-alp-apen-dinar-balk-carp-anatol

Abbreviations: adr – Adriatic, aeg – Aegean, afr – African, alb – Albanian, alp – alpine, alta – Altaic, anatol – Anatolian, apen – Apennine, atl – Atlantic, aral-casp – Aral-Caspian, arc – arctic, balk – Balkan, balt – Baltic, bohem – Bohemian, boratl – Boreal Atlantic, borscand – Boreal-Scandinavian, borrus – Boreal Russian, bor – boreal, borsib – Boreal Siberian, C – central, cant – Cantabrian, carp – Carpathian, catal – Catalonian, cauc – Caucasian, CE – Central European, CEM – Central European mountainous, chin – Chinese, CIRCBOR – Circumboreal, cors – Corsican, cret – Cretan, crim – Crimean, cyp – Cyprian, dac – Dacian, danub – Danubian, daur – Daurian, dinar – Dinaric, E – eastern, EAS – Eurasian, EASM – Eurasian mountainous, eubor – European boreal, eusib – Euro-Siberian, eux – Euxine, gal – Gallic, galec – Galician, hellen – Hellenic, herc – Hertzian, herc.balt – Hertzian Baltic, him – Himalayan, hyrc – Hyrcanian, iber – Iberian, ice – Icelandic, illyr – Illyrian, ion – Ionian, iran – Iranian, jap – Japanese, kam – Kamchatkan, kurd – Kurdish, leb – Lebanese, lusit – Lusitanian, maced – Macedonian, manch – Manchurian, matr – Mátra's, med – Mediterranean, moes – Moesian, mont – montane, N – northern, NAm – North American, NC – north-central, NW – northwestern, or – oriental, pad – Padanian, pann – Pannonian, pelop – Peloponnesian, polon – Polonian, pont – Pontic, provenc – Provencal, pyr – Pyrenean, S – southern, salp – southern alpine, SE – southeastern, SEM – South European mountainous, submerid – submeridonal, sard – Sardinian, sarm – Sarmatian, sib – Siberian, sic – Sicilian; SC – south-central, scand – Scandinavian, sc-pind – Scardo-Pindhian, submed – submediterranean, subalp – subalpine, subbor – Subboreal, subcont – subcontinental, syr – Syrian, sic – Sicilian, subatl – subatlantic, tatr – Tatras, temp – temporal, thrac – Thracian, turan – Turanian, W – western.

In Serbia, *Orchis mascula* subsp. *speciosa* is the only taxon belonging to the Central European-Euxine-Caucasian chorological group (Table 3). This taxon is distributed in Central Europe (including its eastern parts), Southern Europe, and the Balkan Peninsula, west to southeastern France (Vercors) and south to central Italy (Abruzzo) and Corsica (Delforge 2006). *Orchis spitzelii* subsp. *spitzelii* is the only taxon belonging to the South European mountainous chorological group. It is distributed from Sweden (Gotland Island) in the north through the central, southwestern, and southeastern parts of Europe and Western Asia to North Africa (Algeria and Morocco) in the south (Delforge 2006;

WCSP 2020). It is widespread in the Pyrenees, the Alps, the Balkans, Anatolia, and Lebanon (Delforge 2006).

CONCLUSION

Based on personal field investigations, checking and revision of herbarium material, and a review of published sources, the presence of 72 orchid species and subspecies and eight hybrid taxa in Serbia was established. In addition, the presence of three taxa is doubtful. The orchids of Serbia are classified in 22 genera, six tribes, and three subfamilies. The subfamily Orchidoideae is represented by 52 species and subspecies, there are 19 species and subspecies of the subfamily Epidendroideae, while the subfamily Cypripedieae is represented by one species. The most species-rich tribes are Orchideae (49 taxa) and Neottieae (18 taxa). The most species-rich genera are *Epipactis* and *Dactylorhiza* (with 10 taxa each), *Anacamptis* (nine taxa), and *Orchis* (eight taxa). Owing to the great diversity of relief, vegetation, climatic conditions, geological substrates and soils, historical factors, and weak orchid research, it can be assumed that the number of orchid species in Serbia is higher and that new taxa will be recorded in future studies.

The analysis of life forms revealed that tuberous geophytes (51 taxa) are the most numerous, while rhizomatous geophytes are represented by 16 taxa and hemicryptophytes number five orchid taxa. It was also found that most orchids flower in the spring-summer period. Detailed analysis of life forms showed that representatives with ovoid and spindle-shaped tubers are dominant (33 taxa), followed by orchids with rhizomes (21 taxa) and orchids with palmately lobed and fusiform tubers (18 taxa).

Analysis of pollination systems showed that the deceptive system is present in the largest number of orchids (41 taxa), followed by the rewarding pollination system (25 taxa), while the smallest number of orchids have self-pollination (11 taxa). Six orchid species in Serbia are exclusively self-pollinating. The main pollinators of orchids in Serbia are insects from the order Hymenoptera, which pollinate the largest number of orchids, followed by Diptera, Coleoptera, and Lepidoptera, while insects from other insect orders and spiders (Arachnida: Araneidae) are pollinators of a smaller number of orchid taxa.

Phytogeographical analysis shows that the orchids of Serbia belong to nine basic chorological groups. Representatives of the Central European chorological group dominate (with 23 taxa), followed by orchids of the Mediterranean-Submediterranean (16 taxa), Eurasian (12 taxa), and Boreal (nine taxa) groups.

A unique feature of the orchid flora in Serbia is the presence of two Balkan endemics (*Dactylorhiza cordigera* subsp. *bosniaca* and *Himantoglossum calcaratum* subsp. *calcaratum*), as well as three subendemic taxa of the Carpathians and the Balkans (*Dactylorhiza cordigera* subsp. *cordigera*, *Dactylorhiza maculata* subsp. *transsilvanica*, and *Gymnadenia frivaldii*). Orchids whose localities in Serbia represent the northernmost points of their distribution include the following Mediterranean-Submediterranean taxa: *Anacamptis papilionacea*, *Dactylorhiza saccifera*, and *Himantoglossum calcaratum* subsp. *calcaratum*. On the other hand, the following orchids are taxa whose localities in Serbia are the southernmost points of their distribution in this part of Europe: *Dactylorhiza maculata* subsp. *transsilvanica*, *D. maculata* subsp. *maculata*, *D. fuchsii*, and *Traunsteinera globosa*.

Future efforts should be devoted to detailed research on the taxonomy, genetics, and phylogeography of orchids in Serbia, as well as to chorological, ecological, physiological, and conservation studies. Revision of taxa from the groups *Dactylorhiza majalis* and *Dactylorhiza maculata* must be carried out. Special attention should be paid to verification of the occurrence of five orchid taxa for which there are only published data indicating their presence in Serbia: *Anacamptis coriophora* subsp. *fragrans*, *Dactylorhiza romana*, *Neotinea lactea*, *Ophrys sicula*, and *Orchis mascula* subsp. *mascula*. For most endemic and subendemic orchid taxa in Serbia, both pollinator species and pollination mechanisms need to be studied in detail. It would be particularly important to work on the study of morphological and genetic variability, as well as ecophysiological responses of specific orchid taxa from different types of habitats and geological substrates. The influence of biotic factors, i.e., mycorrhizal symbionts and pollinators, on the distribution and abundance of orchids, needs to be studied. Research on the influence of zoo-anthropogenic factors, especially agricultural activities, on orchid performance is also recommended. Finally, there is a need for research on orchid conservation in Serbia (both "*in situ*" and "*ex situ*" conservation of orchids), formation of orchid reserves, and the effectiveness of existing networks for orchid conservation.

ACKNOWLEDGMENTS

This work was supported by the Ministry of Education, Science, and Technological Development of the Republic of Serbia (grant number 451-03-9/2021-14/200178). The authors would like to thank Ivana Stevanoski, Eva Kabaš, Spyros Tsiftsis, Mihai Bobocea, and Aleksandra Maksimović for providing photos of specific orchids. We would also like to thank Dr. Spyros Tsiftsis (Greece) and Dr. Agnieszka Rewicz (Poland) for reviewing the manuscript. We are grateful to Mr. Raymond Dooley for proofreading the manuscript.

REFERENCES

Ackerman, James D. "Diversity and Seasonality of Male Euglossine Bees (Hymenoptera: Apidae) in Central Panamá." *Ecology* 64, no. 2 (1983): 274–83.

Ackerman, James D. "Mechanisms and Evolution of Food-Deceptive Pollination Systems in Orchids." *Lindleyana* 1, no. 2 (1986): 108–13.

Ackerman, James D., and Michael R. Mesler. "Pollination Biology of *Listera cordata* (Orchidaceae)." *American Journal of Botany* 66, no. 7 (1979): 820–24.

Adamović, Lujo. "Beiträge Zur Flora Von Südostserbien." *Österreichische Botanische Zeitschrift* 42 (1892): 404–09. [Contributions to the Flora of Southeast Serbia. *Austrian Botanical Journal*]

Adamović, Lujo. "Die Vegetations formationen Ostserbiens." *Botanische Jahrbücher fur Systematik, Pflanzengeschichte und Pflanzengeographie* 26, no. 2 (1898): 124–218. ["The vegetation formations of Eastern Serbia." *Botanical yearbooks for systematics, plant history and plant geography*]

Adamović, Lujo. "Neue Beiträge zur Flora von Serbien." *Botanisches Centralblatt* 78, no. 10 (1899): 289–97. ["New contributions to the flora of Serbia." *Botanical Centralblatt*]

Adamović, Lujo. "Flora jugoistočne Srbije." *Rad Jugoslavenske akademije znanosti i umjetnosti. Matematičko-prirodoslovni razred* 175, no. 44 (1908): 153–214 [1–62]. [Flora of Southeast Serbia. The work of the Yugoslav Academy of Sciences and Arts. *Mathematics and science class*]

Adamović, Lujo. "Flora jugoistočne Srbije." *Rad Jugoslavenske akademije znanosti i umjetnosti. Matematičko-prirodoslovni razred* 177, no. 45 (1909): 193–240 [63–110]. [Flora of Southeast Serbia. *The work of the Yugoslav Academy of Sciences and Arts. Mathematics and science class*]

Adamović, Lujo. "Građa za floru Kraljevine Crne Gore." *Rad JAZU (Zagreb)* 195 (1913): 1–96. [Material for the flora of the Kingdom of Montenegro. *Work of JAZU (Zagreb)*]

Antonelli, Alexandre, C. Johan Dahlberg, Kaisa H. I. Carlgren, and Thomas Appelqvist. "Pollination of the Lady's Slipper Orchid (*Cypripedium calceolus*) in Scandinavia – Taxonomic and Conservational Aspects." *Nordic Journal of Botany* 27, no. 4 (2009): 266–73.

Antonopoulos, Zissis, and Spyros Tsiftsis. *Atlas of the Greek Orchids, Vol II*. Rethymno, Greece: Mediterraneo Editions, 2017.

Assyov, B., and A. Petrova, eds., *Conspectus of the Bulgarian Vascular Flora: Distribution maps and floristic element*. Fourth edition. Sofia: Bulgarian biodiversity foundation, 2012.

Averyanov, Leonid V. "A Review of the Genus *Dactylorhiza*." In *Orchid Biology: Reviews and Perspectives*, edited by Joseph Arditti, 156–206. V. Oregon: Timber Press, 1990.

Barina, Zoltán, Gabriella Somogyi, Dániel Pifkó, and Marash Rakaj. "Checklist of Vascular Plants of Albania." *Phytotaxa* 378, no. 1 (2018): 1–339.

Bateman, Richard M., Peter M. Hollingsworth, Jillian Preston, Luo Yi-Bo, Alec M. Pridgeon, and Mark W. Chase. "Molecular Phylogenetics and Evolution of Orchidinae and Selected Habenariinae (Orchidaceae)." *Botanical Journal of the Linnean Society* 142, no. 1 (2003): 1–40.

Bateman, Richard M., Attila Molnár V, and Gábor Sramkó. "In Situ Morphometric Survey Elucidates the Evolutionary Systematics of the Eurasian *Himantoglossum* Clade (Orchidaceae: Orchidinae)." *PeerJ* 5 (2017a): e2893.

Bateman, Richard M., Alec M. Pridgeon, and Mark W. Chase. "Phylogenetics of Subtribe Orchidinae (Orchidoideae, Orchidaceae) Based on Nuclear ITS Sequences. 2. Infrageneric Relationships and Taxonomic Revision to Achieve Monophyly of *Orchis* sensu stricto." *Lindleyana* 12 (1997): 113–41.

Bateman, R. M., and P. J. Rudall. "Bumblebee-Mediated Pollination of English Populations of the Military Orchid (*Orchis militaris*): Its Possible Relevance to Functional Morphology, Life History and Climate Change." *New Journal of Botany* 4, no. 3 (2014): 122–33.

Bateman, Richard M., and Paula J. Rudall. "Clarified Relationship Between *Dactylorhiza viridis* and *Dactylorhiza iberica* Renders Obsolete the Former Genus *Coeloglossum* (Orchidaceae: Orchidinae)." *Kew Bulletin* 73, no. 4 (2018). https://doi.org/10.1007/s12225-017-9728-z.

Bateman, R. M., P. J. Rudall, and I. Denholm. "Morphometric Comparison of British *Pseudorchis albida* with Icelandic *P. straminea* (Orchidaceae: Orchidinae)." *New Journal of Botany* 7, no. 2–3 (2017b): 78–93.

Bateman, Richard M., Paula J. Rudall, and Karen E. James. "Phylogenetic Context, Generic Affinities and Evolutionary Origin of the Enigmatic Balkan Orchid *Gymnadenia frivaldii* Hampe Ex Griseb." *Taxon* 55, no. 1 (2006): 107–18.

Baumann, Helmut, Siegfried Künkele, and Richard Lorenz. *Die Orchideen Europas*. Mit angrenzenden Gebieten. Stuttgart: Eugen Ulmer KG, 2006.

Beck-Mannagetta, Günther. "Flora of Bosnia, Herzegovina and Novi Pazar's Sanžak [Flora Bosne, Hercegovine i novopazarskog Sanžaka]" 1(2): *Gymnospermae* i *Monocotyledones*. *Glasnik Zemaljskog muzeja u Bosni i Hercegovini* 1, no 2 (1903): 185–230.

Beck-Mannagetta, Günther. "Flora von Südbosnien und der angrenzenden Hercegovina." Wien: *Annalen des K.K. Naturhistorischen Hofmuseums* 1 (1886): 287–297. ["Flora of South Bosnia and the adjacent Hercegovina." *Vienna: Annals of the K.K. Natural history court museum*]

Bernardos, Sonia, Mónica García-Barriuso, María Ángeles Sánchez-Anta, and Francisco Amich. "Composition, Geographical Affinities and Endemism of the Iberian Peninsula Orchid Flora." *Nordic Journal of Botany* 25, no. 3–4 (2007): 227–37.

Bino, R. J., A. Dafni, and A. D. J. Meeuse. "The Pollination Ecology of *Orchis galilaea* (Bornm. Et Schulze) Schltr. (Orchidaceae)." *New Phytologist* 90, no. 2 (1982): 315–19.

Bonatti, Piera Medeghini, Elisabetta Sgarbi, and Carlo Del Prete. "Gynostemium Micromorphology and Pollination in *Epipactis microphylla* (Orchidaceae)." *Journal of Plant Research* 119, no. 5 (2006): 431–37.

Borbás, Vincze. "Floristicai közlemények a magy. tud. akadémia altal támogatott botanikai kutatásaimból." *Mathematikai és Természettudományi közlemények* 15 (1878): 265–372. ["Floristic announcements are the magy. can. from my botanical research supported by the Academy." *Mathematical and Natural Sciences publications*]

Bornmüller, Joseph. “Beiträge zur Flora Mazedoniens 1.” *Botanische Jahrbücher* 59 (1925): 293–504. [“Contributions to the flora of Macedonia 1.” *Botanical yearbooks*]

Bornmüller, Joseph. “Beiträge zur Flora Mazedoniens III.” *Botanische Jahrbücher fur Systematik, Pflanzengeschichte und Pflanzengeographie* 41, no. 140 (1928): 1–196. [“Contributions to the flora of Macedonia III.” *Botanical yearbooks for systematics, plant history and plant geography*]

Boué, Ami. *La Turquie d’ Europe. Observations sur la géographie, la géologie, l’histoire naturelle, la statistique, les moeurs, les coutumes, l’archéologie, l’agriculture, l’industrie, le commerce, les gouvernements divers, le clergé, l’histoire et l’état politique de cet empire*, Paris: A. Bertrand, 1840. [*Turkey of Europe. Observations on geography, geology, natural history, statistics, manners, customs, archeology, agriculture, industry, commerce, various governments, the clergy, history and political state of this empire*]

Boža, Pal, Vera Budak, and A. Knežević. “Complex *Orchis laxiflora* Lam. subsp. *palustris* (Jacq.) Bonn. et Lay and subsp. *elegans* (Heuff.) Soó at Some Localities in Vojvodina.” *In Proceedings of the scientific conference "Minerals, rocks, extinct and living world of BiH," October 7–8, 1988*, Sarajevo: National Museum of Bosnia and Herzegovina.

Boža, Pal P., and Branislava Butorac. “Contribution to the knowledge of the distribution of some taxa in Vojvodina.” *Bulletin of the Natural History Museum in Belgrade* B36 (1981): 65–74.

Brzosko, Emilia, Andrzej Bajguz, Magdalena Chmur, Justyna Burzyńska, Edyta Jermakowicz, Paweł Mirski, and Piotr Zieliński. “How Are the Flower Structure and Nectar Composition of the Generalistic Orchid *Neottia ovata* Adapted to a Wide Range of Pollinators?” *International Journal of Molecular Sciences* 22, no. 4 (2021): 2214.

Burns-Balogh, Pamela, Dariusz L. Szlachetko, and Amots Dafni. “Evolution, Pollination, and Systematics of the Tribe Neottieae (Orchidaceae).” *Plant Systematics and Evolution* 156, no. 1–2 (1987): 91–115.

Buttler, Karl Peter. *Field Guide to Orchids of Britain and Europe*. Swindon, UK: The Crowood Press, 1991.

Catling, Paul M. "Autogamy in Eastern Canadian Orchidaceae: A Review of Current Knowledge and Some New Observations." *Le Naturaliste Canadien* 110 (1983): 37–53.

Catling, Paul. "Auto-pollination in the Orchidaceae." In *Orchid Biology: Reviews and Perspectives*, Vol. V, edited by Joseph E. Arditti, 121–158. Portland, OR: Timber Press, 1990.

Chase, Mark W., Kenneth M. Cameron, John V. Freudenstein, Alec M. Pridgeon, Gerardo Salazar, Cássio van den Berg, and André Schuiteman. "An Updated Classification of Orchidaceae: Updated Classification of Orchidaceae." *Botanical Journal of the Linnean Society. Linnean Society of London* 177, no. 2 (2015): 151–74.

Chase, Mark, Maarten Christenhusz, and Tom Mirenda. *The Book of Orchids: A Life-Size Guide to Six Hundred Species from Around the World*. London: Ivy Press, 2017.

Christenhusz, Maarten J. M., and James W. Byng. "The Number of Known Plants Species in the World and Its Annual Increase." *Phytotaxa* 261, no. 3 (2016): 201–217.

Claessens, Jean, and Jacques Kleynen. "Column Structure and Pollination of *Corallorhiza trifida* Châtelain (Orchidaceae)." *Journal Europäischer Orchideen* 30 (1998): 629–37.

Claessens, Jean, and Jacques Kleynen. "Many Ways to Get Happy: Pollination Modes of European *Epipactis* species." *Orchid Digest* 80, no. 3 (2016a): 144–52.

Claessens, Jean, and Jacques Kleynen. *The Flower of the European Orchid: Form and Function*. Geulle, The Netherlands: Self Published, 2011.

Claessens, Jean, and Jacques Kleynen. "The Pollination of European Orchids: Part 2: *Cypripedium* and *Cephalanthera*." *Journal of the Hardy Orchid Society* 10, no. 4(70) (2013): 114–20.

Claessens, Jean, and Jacques Kleynen. "The Pollination of European Orchids Part 4: *Goodyera* and *Spiranthes*." *Journal of the Hardy Orchid Society* 13 (2016b): 54–61.

Claessens, Jean, and Jacques Kleynen. "The Pollination of European Orchids Part 5: *Himantoglossum* and *Anacamptis*, Two Examples of Deceptive Pollination." *Journal of the Hardy Orchid Society* 13, no. 4 (2016c): 114–22.

Claessens, Jean, and Jacques Kleynen. " The Pollination of European Orchids Part 3: *Limodorum* and *Epipactis*." *Journal of the Hardy Orchid Society* 11, no. 2 (2014): 64–71.

Claessens, Jean, and Bernhard Seifert. "Ant Pollination of *Dactylorhiza viridis*." *Orchid Digest* 82, no. 3 (2018): 154–58.

Claessens, Jean, and Bernhard Seifert. "Significant Ant Pollination in Two Orchid Species in the Alps as Adaptation to the Climate of the Alpine Zone?" *Tuexenia* 37 (2017): 363–74.

Cozzolino, Salvatore, Florian P. Schiestl, Andreas Müller, Olga De Castro, Antonio Marco Nardella, and Alex Widmer. "Evidence for Pollinator Sharing in Mediterranean Nectar-Mimic Orchids: Absence of Premating Barriers?" *Proceedings of the Royal Society B: Biological Sciences* 272, no. 1569 (2005): 1271–78.

Cozzolino, Salvatore, and Alex Widmer. "Orchid Diversity: An Evolutionary Consequence of Deception?" *Trends in Ecology & Evolution* 20, no. 9 (2005): 487–94.

Čolović-Parabućski, Stanija. "Orchids of Fruška Gora [Kaćuni Fruške gore]." *Zaštita prirode* 12 (1958): 31–36.

Dafni, Amots. "Pollination in *Orchis* and related genera: evolution from reward to deception." In *Orchid biology, reviews and perspectives* IV, edited by Joseph Arditti, 79–104. Ithaca: Cornell University Press, 1987.

Dafni, A., and Y. Ivri. "Pollination Ecology of, and Hybridization Between, *Orchis coriophora* L. and *O. collina* Sol. ex Russ. (Orchidaceae) in Israel." *New Phytologist* 83, no. 1 (1979): 181–87.

Dafni, Amots, and Yariv Ivri. "The Flower Biology of *Cephalanthera longifolia* (Orchidaceae) – Pollen Imitation and Facultative Floral Mimicry." *Plant Systematics and Evolution* 137, no. 4 (1981): 229–40.

Dafni, Amots, Yariv Ivri, and N. B. M. Brantjes. "Pollination of *Serapias vomeracea* Briq. (Orchidaceae) by Imitation of Holes for Sleeping Solitary Male Bees (Hymenoptera)." *Acta Botanica Neerlandica* 30, no. 1–2 (1981): 69–73.

De Angelli, Nora, and Dan Anghelescu. *Orchids of Romania,* Snagov, Romania, 2020.

Delforge, Pierre. *Orchids of Europe, North Africa and the Middle East.* London: A. & C. Black, 2006.

Devos, Nicolas, Olivier Raspé, Anne–laure Jacquemart, and Daniel Tyteca. "On the Monophyly of *Dactylorhiza* Necker Ex Nevski (Orchidaceae): Is *Coeloglossum viride* (L.) Hartman a *Dactylorhiza*?" *Botanical Journal of the Linnean Society* 152, no. 3 (2006): 261–69.

Diklić, Nikola. "Fam. Orchidaceae." In *Flora SR Srbije VIII,*" edited by Mladen Josifović, 36–116. Belgrade: Serbian Academy of Sciences and Arts, 1976.

Djordjević, Vladan. "*Epipactis muelleri* (Orchidaceae-Neottieae), A Species New to the Flora of Serbia." *Phyton (Horn)* 56, no. 2 (2016): 303–12.

Djordjević, Vladan. *Orchids of Mt. Maljen (Western Serbia): Ecology, Distribution and Conservation.* Diploma thesis, University of Belgrade, 2012.

Djordjević, Vladan. "Orchids of the Valjevo Region [Orhideje Valjevskog kraja]." *Petničke sveske* 42 (1997): 3–17.

Djordjević, Vladan. *Spatial Distribution and Ecology of Orchids (Orchidaceae) of Western Serbia.* PhD diss., University of Belgrade, 2018.

Djordjević, Vladan, Ksenija Jakovljević, and Vladimir Stevanović. "Three Taxa of *Epipactis* (Orchidaceae-Epidendroideae) New For the Flora of Serbia." *Phyton (Horn)* 56, no. 1 (2016a): 77–89.

Djordjević, Vladan, Slobodan Jovanović, and Vladimir Stevanović, "*Dactylorhiza fuchsii* (Orchidaceae), a New Species in the Flora of Serbia." *Archives of Biological Sciences* 66, no. 3 (2014): 1227–32.

Djordjević, Vladan, Dmitar Lakušić, Slobodan Jovanović, and Vladimir Stevanović. "Distribution and Conservation Status of Some Rare and Threatened Orchid Taxa in the Central Balkans and the Southern Part of the Pannonian Plain." *Wulfenia* 24 (2017): 143–62.

Djordjević, Vladan, Marjan Niketić, and Gordana Tomović. "Liliopsida: Orchidaceae." In *An annotated Checklist of Vascular Flora of Serbia 1. Lycopodiopsida, Polypodiopsida, Gnetopsida, Pinopsida and Liliopsida [Catalogue of taxa]*, Marjan Niketić, and Gordana Tomović, 102–110. Belgrade: Serbian Academy of Sciences and Arts, Monographs 690, Department of Chemical and Biological Sciences 10, 2018.

Djordjević, V., Gordana Tomović, and D. Lakušić. "*Epipactis purpurata* Sm. (Orchidaceae): A New Species in the Flora of Serbia." *Archives of Biological Sciences* 62, no. 4 (2010): 1175–79.

Djordjević, Vladan, and Spyros Tsiftsis. "Patterns of Orchid Species Richness and Composition in Relation to Geological Substrates." *Wulfenia* 26 (2019): 1–21.

Djordjević, Vladan, and Spyros Tsiftsis. "The Role of Ecological Factors in Distribution and Abundance of Terrestrial Orchids." In *Orchids Phytochemistry, Biology and Horticulture, Reference Series in Phytochemistry. Fundamentals and Applications*, edited by Jean-Michel Merillon, and Hippolyte Kodja, 1–71. Cham: Springer International Publishing, 2020, doi:10.1007/978-3-030-11257-8_4-1.

Djordjević, Vladan, Spyros Tsiftsis, Ksenija Jakovljević, Jasmina Šinžar-Sekulić, and Snežana Vukojičić. "First Record of a Natural Hybrid *Neotinea* × *dietrichiana* (Orchidaceae) in Serbia." *Phytologia Balcanica* 18 (2012): 163–71.

Djordjević, Vladan, Spyros Tsiftsis, Dmitar Lakušić, Slobodan Jovanović, Ksenija Jakovljević, and Vladimir Stevanović. "Patterns of Distribution, Abundance and Composition of Forest Terrestrial Orchids." *Biodiversity and Conservation* 29, no. 14 (2020a): 4111–34.

Djordjević, Vladan, Spyros Tsiftsis, Dmitar Lakušić, Slobodan Jovanović, and Vladimir Stevanović. "Factors Affecting the

Distribution and Abundance of Orchids in Grasslands and Herbaceous Wetlands." *Systematics and Biodiversity* 14, no. 4 (2016b): 355–70.

Djordjević, V., S. Tsiftsis, D. Lakušić, and V. Stevanović. "Niche Analysis of Orchids of Serpentine and Non-Serpentine Areas: Implications for Conservation." *Plant Biosystems* 150, no. 4 (2016c): 710–19.

Djordjević, Vladan, Spyros Tsiftsis, Dmitar Lakušić, Slobodan Jovanović, and Vladimir Stevanović. "Orchid Species Richness and Composition in Relation to Vegetation Types." *Wulfenia* 27 (2020b): 183–210.

Djordjević, Vladan, Milorad Veljković, and Snežana Vukojičić. "*Orchis provincialis*, a New Species in the Flora of Serbia." *Wulfenia* 28 (2021): in press.

Dolinar, Branko. *Orchids in Slovenia* [*Kukavičke v Sloveniji*]. Ljubljana: Pipinova knjiga, 2015.

Dötterl, S., and N. J. Vereecken. "The Chemical Ecology and Evolution of Bee–Flower Interactions: A Review and Perspectives." *Canadian Journal of Zoology* 88, no. 7 (2010): 668–97.

Dressler, Robert L. *The Orchids: Natural History and Classification*. Cambridge, Mass: Horvard University Press, 1981.

Dulić, Jovana. *Biodiversity and Breeding of Terrestrial Orchids from Fruška Gora Mountain*. PhD diss., University of Novi Sad, 2019.

Efimov, Petr G. "A Revision of *Platanthera* (Orchidaceae; Orchidoideae; Orchideae) in Asia." *Phytotaxa* 254, no. 1 (2016): 1–233.

Efimov, Peter. "Notes on *Epipactis condensata*, *E. rechingeri* and *E. purpurata* (Orchidaceae) in the Caucasus and Crimea." *Willdenowia* 38, no. 1 (2008): 71–80.

Ellenberg, H., and D. Mueller-Dambois. "A Key to Raunkiaer Plant Life Forms with Revised Subdivisions." ETH, Zurich: *Berichte des Geobotanischen Institutes der Eidg. Techn. Hochschule, Stiftung Rübel* 37 (1967): 56–73.

Esposito, Fabiana, Nicolas J. Vereecken, Maddalena Gammella, Rosita Rinaldi, Pascal Laurent, and Daniel Tyteca. "Characterization of

Sympatric *Platanthera bifolia* and *Platanthera chlorantha* (Orchidaceae) Populations with Intermediate Plants." *PeerJ* 6 (2018): e4256.

Faegri, K., and L. van Der Pijl. *The Principles of Pollination Ecology*. Oxford: Pergamon Press, 1979.

Fantinato, Edy, Silvia Del Vecchio, Mattia Baltieri, Beatrice Fabris, and Gabriella Buffa. "Are Food-Deceptive Orchid Species Really Functionally Specialized for Pollinators?" *Ecological Research* 32, no. 6 (2017): 951–59.

Farrell, L. "Biological Flora of the British Isles: *Orchis militaris* L." *Journal of Ecology* 73, no. 3 (1985): 1041–53.

Fay, Michael F., Brendan Sayers, and Ian Taylor. "802. *Dactylorhiza viridis*: (Orchidaceae)" *Curtis's Botanical Magazine* 32, no. 1 (2015): 33–41.

Fekete, Réka, Timea Nagy, Judit Bódis, Éva Biró, Viktor Löki, Kristóf Süveges, Attila Takács, Jácint Tökölyi, and Attila Molnár V. "Roadside Verges as Habitats for Endangered Lizard-Orchids (*Himantoglossum* spp.): Ecological Traps or Refuges?" *Science of the Total Environment* 607–608 (2017): 1001–8.

Fenster, Charles B., and Silvana Martén-Rodríguez. "Reproductive Assurance and the Evolution of Pollination Specialization." *International Journal of Plant Sciences* 168, no. 2 (2007): 215–28.

Fritsch, Karl. "Neue Beiträge zur Flora der Balkanhalbinsel, insbensondere Serbiens, Bosniens und der Hercegowina I." *Mitt. Naturwiss. Vereines Steiermark* 45, no. 1 (1909): 131–183. ["New contributions to the flora of the Balkan Peninsula, in particular Serbia, Bosnia and Herzegovina I." *Mitt. Naturwiss. Association of Styria*]

Fritz, Anna-Lena. "Deceit Pollination of *Orchis spitzelii* (Orchidaceae) on the Island of Gotland in the Baltic: A Suboptimal System." *Nordic Journal of Botany* 9, no. 6 (1990): 577–87.

Fritz, Anna-Lena, and L. Anders Nilsson. "Reproductive Success and Gender Variation in Deceit-Pollinated Orchids." In *Floral Biology. Studies on Floral Evolution in Animal-Pollinated Plants*, edited by

David Lloyd, G., and Spencer Barret, C. H., 319–338. New York: Chapman and Hall, 1996.

Gajić, Milovan. *Flora and Vegetation of Golija and Javor* [*Flora i vegetacija Golije i Javora*]. Beograd: Šumarski fakultet; Ivanjica: OOUR Šumarstvo "Golija," 1989.

Gajić, Milovan, Momčilo Kojić, Dragan Karadžić, Miloje Vasiljević, and Mihailo Stanić. *Vegetation of Tara National Park* [*Vegetacija Nacionalnog parka Tara*]. Beograd: Šumarski fakultet; Bajina Bašta: Nacionalni park Tara, 1992.

Gaskett, A. C. "Orchid Pollination by Sexual Deception: Pollinator Perspectives." *Biological Review* 86, no. 1 (2011): 33–75.

Gervasi, Daniel D. L., Marc-Andre Selosse, Mathieu Sauve, Wittko Francke, Nicolas J. Vereecken, Salvatore Cozzolino, and Florian P. Schiestl. "Floral Scent and Species Divergence in a Pair of Sexually Deceptive Orchids." *Ecology and Evolution* 7, no. 15 (2017): 6023–34.

Girlanda, M., M. A. Selosse, D. Cafasso, F. Brilli, S. Delfine, R. Fabbian, S. Ghignone, et al. "Inefficient Photosynthesis in the Mediterranean Orchid *Limodorum abortivum* Is Mirrored by Specific Association to Ectomycorrhizal Russulaceae" *Molecular Ecology* 15, no. 2 (2006): 491–504.

GIROS. *Orchidee d'Italia – Guida alle orchidee spontanee*. Cornaredo (Milano): Il Castello, 2009.

Givnish, Thomas J., Daniel Spalink, Mercedes Ames, Stephanie P. Lyon, Steven J. Hunter, Alejandro Zuluaga, Alfonso Doucette, et al. "Orchid Historical Biogeography, Diversification, Antarctica and the Paradox of Orchid Dispersal." *Journal of Biogeography* 43, no. 10 (2016): 1905–16.

Givnish, Thomas J., Daniel Spalink, Mercedes Ames, Stephanie P. Lyon, Steven J. Hunter, Alejandro Zuluaga, William J. D. Iles, et al. "Orchid Phylogenomics and Multiple Drivers of Their Extraordinary Diversification." *Proceedings of the Royal Society B: Biological Sciences* 282, no. 1814 (2015): 20151553.

Gladkova, N. V. “The family Orchidaceae.” In *Zhizn’ rastenii* [*Plant life*], vol. 6, edited by A. L. Takhtajan 248–275. Moscow: Prosveshchenie, 1982.

Godra, Bartholomäus. “Flora des Peterwardeiner Grenz – Regiments Nr. 9.” *Österreichische botanische Zeitschrift* 22 (1872): 98–100, 133–135, 163–166, 201–206, 228–231, 266–270, 292–297. [“Flora of the Peterwardeiner Grenz - Regiment No. 9.” *Austrian botanical journal*]

Gravendeel, Barbara, Ann Smithson, Ferry J. W. Slik, and Andre Schuiteman. “Epiphytism and Pollinator Specialization: Drivers for Orchid Diversity?” *Philosophical Transactions of the Royal Society of London. Series B, Biological Sciences* 359, no. 1450 (2004): 1523–35.

Grisebach, August. *Spicilegium Florae rumelicae et bithynicae exhibens synopsin plantarum quas in aest. 1839 legit auctor A. Grisebach* 2, 161–548. Fridericus Vieweg et filius, Braunschweig [“Brunsvigae”], 1846. [*Spicilegium Flora rumeicae and bithynicae exhibiting a synopsis of plants in est. 1839 author A. Grisebach reads*]

Gustafsson, A. Lovisa S., Christiano F. Verola, and Alexandre Antonelli. “Reassessing the Temporal Evolution of Orchids with New Fossils and a Bayesian Relaxed Clock, with Implications for the Diversification of the Rare South American Genus *Hoffmannseggella* (Orchidaceae: Epidendroideae).” *BMC Evolutionary Biology* 10 (2010): 177.

Gutowski, Jerzy M. “Pollination of the Orchid *Dactylorhiza fuchsii* by Longhorn Beetles in Primeval Forests of Northeastern Poland.” *Biological Conservation* 51, no. 4 (1990): 287–97.

Hágsater, Eric, and Vinciane Dumont, V., eds. *Orchids: Status, Survey and Conservation Action Plan*. Gland: IUCN, 1996.

Haraštová-Sobotková, Milena, Jana Jersáková, Pavel Kindlmann, and Ladislav Čurn. “Morphometric and Genetic Divergence among Populations of *Neotinea ustulata* (Orchidaceae) with Different Flowering Phenologies.” *Folia Geobotanica* 40, no. 4 (2005): 385–405.

Hausmann, Axel. *The Geometrid Moths of Europe 2*. Sterrhinae. Stenstrup: Apollo Books, 2004.

Hayek, August von. "Prodromus Florae Peninsulae Balcanicae 3." Berlin–Dahlem: *Repertorium specierum novarum regni vegetabilis* 30 (1933): 371–416.

Hedrén, Mikael, Erich Klein, and Herwig Teppner. "Evolution of Polyploids in the European Orchid Genus *Nigritella*: Evidence from Allozyme Data." *Phyton (Horn)* 40, no. 2 (2000): 239–75.

Hedrén, Mikael, Richard Lorenz, and David Ståhlberg. "Evidence for Bidirectional Hybridization Between *Gymnadenia* and *Nigritella.* " *Journal Europäischer Orchideen* 50, no. 1 (2018a): 43–60.

Hedrén, Mikael, Richard Lorenz, Herwig Teppner, Branko Dolinar, Cesario Giotta, Norbert Griebl, Sven Hansson, et al. "Evolution and Systematics of Polyploid *Nigritella* (Orchidaceae)." *Nordic Journal of Botany* 36, no. 3 (2018b): njb-01539.

Heinrich, Bernd. "Bee Flowers: A Hypothesis on Flower Variety and Blooming Times." *Evolution* 29, no. 2 (1975): 325–334.

Hill, David A. "Orchids in Southern Parts of Serbia: Some Recent Findings." *Niš: Centenary symposium of the 'The Flora around Niš'*(1985): 41–43.

Huber, Franz K., Roman Kaiser, Willi Sauter, and Florian P. Schiestl. "Floral Scent Emission and Pollinator Attraction in Two Species of *Gymnadenia* (Orchidaceae)." *Oecologia* 142, no. 4 (2005): 564–75.

Inda, Luis A., Manuel Pimentel, and Mark W. Chase. "Phylogenetics of Tribe *Orchideae* (Orchidaceae: Orchidoideae) Based on Combined DNA Matrices: Inferences Regarding Timing of Diversification and Evolution of Pollination Syndromes." *Annals of Botany* 110, no. 1 (2012): 71–90.

Jacquemyn, Hans, Rein Brys, Martin Hermy, and Jo H. Willems. "Does Nectar Reward Affect Rarity and Extinction Probabilities of Orchid Species? An Assessment Using Historical Records from Belgium and the Netherlands." *Biological Conservation* 121, no. 2 (2005a): 257–63.

Jacquemyn, Hans, Rein Brys, Olivier Honnay, and Michael J. Hutchings. "Biological Flora of the British Isles: *Orchis mascula* (L.) L." *Journal of Ecology* 97, no. 2 (2009): 360–77.

Jacquemyn, Hans, Rein Brys, and Michael J. Hutchings. "Biological Flora of the British Isles: *Epipactis palustris*." *Journal of Ecology* 102, no. 5 (2014): 1341–55.

Jacquemyn, Hans, and Michael J. Hutchings. "Biological Flora of the British Isles: *Spiranthes spiralis* (L.) Chevall." *Journal of Ecology* 98, no. 5 (2010): 1253–67.

Jacquemyn, Hans, Claire Micheneau, David L. Roberts, and Thierry Pailler. "Elevational Gradients of Species Diversity, Breeding System and Floral Traits of Orchid Species on Reunion Island." *Journal of Biogeography* 32, no. 10 (2005b): 1751–61.

Jakovljević, Ksenija, Gordana Tomović, Vladan Djordjević, Marjan Niketić, and Vladimir Stevanović. "Steppe Flora in Serbia – Distribution, Ecology, Centres of Diversity and Conservation Status." *Folia Geobotanica* 55, no. 1 (2020): 1–14.

Jakubska-Busse, Anna, and Marcin Kadej. "The Pollination of *Epipactis* Zinn, 1757 (Orchidaceae) Species in Central Europe – the Significance of Chemical Attractants, Floral Morphology and Concomitant Insects." *Acta Societatis Botanicorum Poloniae* 80, no. 1 (2011): 49–57.

Jakubska-Busse, Anna, Spyros Tsiftsis, Michał Śliwiński, Zdenka Křenová, Vladan Djordjević, Corina Steiu, Marta Kolanowska, et al. "How to Protect Natural Habitats of Rare Terrestrial Orchids Effectively: A Comparative Case Study of *Cypripedium calceolus* in Different Geographical Regions of Europe." *Plants* 10, no. 2 (2021): 404.

Jersáková, Jana, Steven D. Johnson, and Pavel Kindlmann. "Mechanisms and Evolution of Deceptive Pollination in Orchids." *Biological Reviews* 81, no. 02 (2006): 219–35.

Jersáková, Jana, Tamara Malinová, Kateřina Jeřábková, and Stefan Dötterl. "Biological Flora of the British Isles: *Pseudorchis albida* (L.) Á. & D. Löve." *Journal of Ecology* 99, no. 5 (2011): 1282–98.

Jersáková, Jana, Johannes Spaethe, Martin Streinzer, Johann Neumayer, Hannes Paulus, Stefan Dötterl, and Steven D. Johnson. "Does *Traunsteinera globosa* (the Globe Orchid) Dupe Its Pollinators through Generalized Food Deception or Mimicry?" *Botanical Journal of the Linnean Society* 180, no. 2 (2016): 269–94.

Jersáková, Jana, Iva Traxmandlová, Zdeněk Ipser, Matthias Kropf, Giuseppe Pellegrino, Bertrand Schatz, Vladan Djordjević, Pavel Kindlmann, and Susanne S. Renner. "Biological Flora of Central Europe: *Dactylorhiza sambucina* (L.) Soó." *Perspectives in Plant Ecology, Evolution and Systematics* 17, no. 4 (2015): 318–29.

Jin, Wei-Tao, André Schuiteman, Mark W. Chase, Jian-Wu Li, Shih-Wen Chung, Tian-Chuan Hsu, and Xiao-Hua Jin. "Phylogenetics of Subtribe Orchidinae s.l. (Orchidaceae; Orchidoideae) Based on Seven Markers (Plastid MatK, PsaB, RbcL, TrnL-F, TrnH-Psba, and Nuclear NrITS, Xdh): Implications for Generic Delimitation." *BMC Plant Biology* 17, no. 1 (2017): 222.

Johnson, S. D., and L. A. Nilsson. "Pollen Carryover, Geitonogamy, and the Evolution of Deceptive Pollination Systems in Orchids." *Ecology* 80, no. 8 (1999): 2607–19.

Jurišić, Živojin J. "News from Science and Teaching – News from Tornik [Novine iz nauke i nastave – Novina sa Tornika]." In *Nastavnik - list profesorskoga društva*, edited by J. Živanović, 187–88. Belgrade: Nova štamparija Davidović, 1913. [*Teacher - a journal of professor society*]

Jurišić, Živojin. "A Contribution to the Flora of Southern Serbia [Prilog flori južne Srbije]." *Spomenik Srpske Kraljevske Akademije LX, Prvi razred* 10 (1923): 3–45. [*Monument of the Serbian Royal Academy LX, First grade*]

Jurišić, Živojin. "Novelties for the Flora of the Kingdom of Serbia [Prinove za floru Kraljevine Srbije]." *Prosvetni glasnik* 22, no. 7 (1901): 857–66. [*An educational bulletin*]

Karadžić, B., Žaklina Marjanović, A. Mijović, S. Marinković, and Ranka Popović. "*Ophrys insectifera* L., A New Orchid Species in Flora of Serbia." *Archives of Biological Science* 52 (2000): 11–12.

Košanin, Nedeljko. "Elements of Vlasina Flora [Elementi vlasinske flore]." Beograd: *Muzej srpske zemlje* 10 (1910): 1–42. [*Museum of Serbian land*]

Kotilínek, Milan, Irina Tatarenko, and Jana Jersáková. "Biological Flora of the British Isles: *Neottia cordata*." *Journal of Ecology* 106, no. 1 (2018): 444–60.

Kotilínek, Milan, Tamara Těšitelová, and Jana Jersáková. "Biological Flora of the British Isles: *Neottia ovata*." *Journal of Ecology* 103, no. 5 (2015): 1354–66.

Kowalkowska, Agnieszka K., Hanna B. Margońska, Małgorzata Kozieradzka-Kiszkurno, and Jerzy Bohdanowicz. "Studies on the Ultrastructure of a Three-Spurred Fumeauxiana Form of *Anacamptis pyramidalis*." *Plant Systematics and Evolution* 298, no. 6 (2012): 1025–35.

Kretzschmar, Horst, Wolfgang Eccarius, and Helga Dietrich. *The Orchid Genera Anacamptis*, *Orchis* and *Neotinea*. *Phylogeny, Taxonomy, Morphology, Biology, Distribution, Ecology and Hybridization*. 2nd ed. Bürgel: EchinoMedia Verlag, 2007.

Kreutz, Karel C. A. J., and Çolak, Alper Hüseyin, eds. *Türkiye Orkideleri*. Istanbul: Rota Yayınlari, 2009. [*Turkey Orchids*.]

Kull, Tiiu, and Michael J. Hutchings. "A Comparative Analysis of Decline in the Distribution Ranges of Orchid Species in Estonia and the United Kingdom." *Biological Conservation* 129, no. 1 (2006): 31–39.

Lakušić, Dmitar. *High Mountain Flora of Mt Kopaonik – Ecological-Phytogeographical Study*. Master's Thesis, University of Belgrade, 1993.

Lang, David. *Britain's Orchids, a Guide to the Identification and Ecology of the Wild Orchids of Britain and Ireland*. Old Basing, Hampshire: Wild Guides Ltd., 2004.

Lind, Henrik, Markus Franzén, Börge Pettersson†, and L. Anders Nilsson. "Metapopulation Pollination in the Deceptive Orchid *Anacamptis pyramidalis*." *Nordic Journal of Botany* 25, no. 3–4 (2007): 176–82.

Li, Taiqiang, Shimao Wu, Wenke Yang, Marc-André Selosse, and Jiangyun Gao. "How Mycorrhizal Associations Influence Orchid Distribution and Population Dynamics." *Frontiers in Plant Science* 12 (2021): 647114.

Majstorović, Ilija S. *Floristic Observations of the Surroundings of Valjevo* [*Floristička zapažanja okoline Valjeva*] *I*. Valjevo: Štamparija Tih. M. Petković, 1929.

Markgraf, Friedrich. "Pflanzengeographie von Albanien. Ihre Bedeutung für Vegetation und Flora der Mittelmeerländer." *Bibliotheca Botanica* 26 (1932): 6–132. [Plant Geography of Albania. Their importance for the vegetation and flora of the Mediterranean countries.]

McKendrick, S. L., J. R. Leake, D. Lee Taylor, and D. J. Read. "Symbiotic Germination and Development of the Myco-heterotrophic Orchid *Neottia nidus-avis* in Nature and Its Requirement for Locally Distributed Sebacina Spp." *New Phytologist* 154, no. 1 (2002): 233–47.

Meekers, Tine, Michael J. Hutchings, Olivier Honnay, and Hans Jacquemyn. "Biological Flora of the British Isles: *Gymnadenia conopsea* s.l." *Journal of Ecology* 100, no. 5 (2012): 1269–88.

Meikle, R. D. *Flora of Cyprus*, vol. *2*. Bentham Moxon Trust. Kew: Royal Botanic Gardens, 1985.

Meusel, Hermann, Eckehardt Jäger, and Erich Weinert. *Vergleichende Chorologie der Zentraleuropäischen Flora 1*. Text. Karten. Jena: Gustav Fischer, 1965. [*Comparative chorology of the Central European flora 1*.]

Meusel, Hermann, Eckehardt Jäger, and Erich Weinert. *Vergleichende Chorologie der Zentraleuropäischen Flora 2*. Text. Karten. Jena: Gustav Fischer, 1978. [*Comparative chorology of the Central European flora 2*.]

Meusel, Hermann, and Eckehardt Jäger. *Vergleichende Chorologie der zentraleuropäischen Flora 3*. Karten, Text, Literatur, Register. Jena, Stuttgart: Gustav Fischer, 1992. [*Comparative chorology of the Central European flora 3*.]

Micheneau, Claire, Jacques Fournel, and Thierry Pailler. "Bird Pollination in an Angraecoid Orchid on Reunion Island (Mascarene Archipelago, Indian Ocean)." *Annals of Botany* 97, no. 6 (2006): 965–74.

Molnár, Attila, ed. *Magyarország orchideáinak atlasza*. Budapest: Kossuth kiadó, 2011. [*Atlas of orchids in Hungary*.]

Molnár, Attila, Karel C. A. J. Kreutz, Miklós Óvári, Alexander N. Sennikov, Richard M. Bateman, Attila Takács, Lajos Somlyay, and Gábor Sramkó. "*Himantoglossum jankae* (Orchidaceae: Orchideae), a New Name for a Long-Misnamed Lizard Orchid." *Phytotaxa* 73, no. 1 (2012): 8–12.

Moore, D. M. "CCIII. Orchidaceae." In *Flora Europaea 5*, edited by T. G. Tutin, V. H. Heywood, N. A. Burges, D. M. Moore, D. H. Valentine, S. M. Walters, and D. A. Webb, 325–350. Cambridge: Camridge University Press, 1980.

Nahirnić, Ana, and Stoyan Beshkov. "Novelties in the Zygaeninae fauna (Lepidoptera: Zygaenidae) of Albania and the Republic of Macedonia." *Acta Zoologica Bulgarica* 70, no. 4 (2018): 465–468.

Nazarov, Vladimir, and Ulf Buchsbaum. "Widderchen als Bestäuber von Orchideen in Thüringen (Insecta: Lepidoptera, Zygaenidae)." *Entomofauna* 25, no. 25 (2004): 365–72. [Ram as a pollinator of orchids in Thuringia (Insecta: Lepidoptera, Zygaenidae).]

Nazarov, Vladimir V., and Konstantin Efetov, A. "On the role of the Zygaenidae (Lepidoptera) in pollination of *Anacamptis pyramidalis* (Orchidaceae)." *Zoological Journal* 72 (1993): 54–67.

Ničić, Đorđe. "Material for the Flora of the Vranje Area [Građa za floru okoline Vranje]." *Nastavnik* 5, no. 1–6 (1894): 33–78.

Niketić, Marjan, and Gordana Tomović. *An Annotated Checklist of Vascular Flora of Serbia 1. Lycopodiopsida, Polypodiopsida, Gnetopsida, Pinopsida and Liliopsida*. Belgrade: Serbian Academy of Sciences and Arts, 2018.

Niketić, Marjan, Gordana Tomović, Goran Anačkov, Mirjana Bartula, Sladjana Djordjević, Vladan Djordjević, Suzana Djordjević-Milošević, Šemija Duraki, Milan Gavrilović, Peđa Janaćković, Eva

Kabaš, Nevena Kuzmanović, Dmitar Lakušić, Predrag Lazarević, Ranko Perić, Vladimir Randjelović, Dragiša Savić, Mihajlo Stanković, Ivana Stevanoski, Verica Stojanović, Olga Vasić, Snežana Vukojičić, Bojan Zlatković, and Vladimir Stevanović. "Material on the Annotated Checklist of Vascular Flora of Serbia. Nomenclatural, Taxonomic and Floristic Notes *II*." *Bulletin of the Natural History Museum in Belgrade* 13 (2020): 87–169.

Niketić, Marjan, Gordana Tomović, Ranko Perić, Bojan Zlatković, Goran Anačkov, Vladan Djordjević, Nejc Jogan, Boris Radak, Šemija Duraki, Mihajlo Stanković, Nevena Kuzmanović, Dmitar Lakušić, and Vladimir Stevanović. "Material on the Annotated Checklist of Vascular Flora of Serbia. Nomenclatural, Taxonomic and Floristic Notes I." *Bulletin of the Natural History Museum in Belgrade* 11 (2018): 101–180.

Nikolić, Toni. *Flora Croatica Vol. 4: Vaskularna flora Republike Hrvatske, Ekskurzijska flora*, Zagreb: Alfa d.d., 2019. [*Flora Croatica Vol. 4: Vascular flora of the Republic of Croatia, Excursion flora*]

Nilsson, L. Anders. "Pollination Ecology and Evolutionary Process in Six Species of Orchids." *Abstracts of Uppsala Dissertations from the Faculty of Science and Technology* 59 (1981a): 1–40.

Nilsson, L. Anders. "Anthecology of *Orchis mascula* (Orchidaceae)." *Nordic Journal of Botany* 3, no. 2 (1983a): 157–79.

Nilsson, L. Anders. "Mimesis of Bellflower (*Campanula*) by the Red Helleborine Orchid *Cephalanthera rubra*." *Nature* 305, no. 5937 (1983b): 799–800.

Nilsson, L. Anders. "Orchid Pollination Biology." *Trends in Ecology & Evolution* 7, no. 8 (1992): 255–59.

Nilsson, L. Anders. "Pollination Ecology of *Epipactis palustris* (Orchidaceae)." *Botaniska Notiser* 131 (1978): 355–68.

Nilsson, L. Anders. "Pollination ecology of *Herminium monorchis* (Orchidaceae)." *Botaniska Notiser* 132 (1979): 537–49.

Nilsson, L. Anders. "Processes of Isolation and Introgressive Interplay between *Platanthera bifolia* (L.) Rich and *P. chlorantha* (Custer)

Reichb. (Orchidaceae)." *Botanical Journal of the Linnean Society* 87, no. 4 (1983c): 325–50.

Nilsson, L. Anders. "The Pollination Ecology of *Dactylorhiza sambucina* (Orchidaceae)." *Botaniska Notiser* 133 (1980): 367–85.

Nilsson, L. Anders. "The Pollination Ecology of *Listera ovata* (Orchidaceae)," *Nordic Journal of Botany* 1, no. 4 (1981b): 461–80.

Novák, Frant A. "Ad florae Serbiae cognitionem additamentum primum." *Preslia* 4 (1926): 37–56.

Obradović, Melanija, Vera Budak, and Pal Boža. "New Infraspecies Taxa of Orchids (Orchidaceae Lindl.) in the Flora of Vojvodina." *Matica Srpska* 64 (1983): 119–130.

Ostrowiecka, Beata, Izabela Tałałaj, Emilia Brzosko, Edyta Jermakowicz, Paweł Mirski, Agata Kostro-Ambroziak, Łukasz Mielczarek, et al. "Pollinators and Visitors of the Generalized Food-Deceptive Orchid *Dactylorhiza majalis* in North-Eastern Poland." *Biologia* 74, no. 10 (2019): 1247–57.

Pančić, Josif. "Verzeichniss der in Serbien wildwachsenden Phanerogamen, nebst den Diagnosen eineger neuer Arten." *Verhandlungen der kaiserlich–königlichen zoologisch–botanischen Gesellschaft in Wien* 6 (1856): 475–598. [Directory of the wild phanerogams growing in Serbia, together with the diagnoses of some new species. *Negotiations of the Imperial-Royal Zoological-Botanical Society in Vienna*]

Pančić, Josif. *Flora in the Environs of Belgrade: According to the Analytical Method Compiled by Dr. Josif Pančić* [*Flora u okolini beogradskoj. Po analitičkom metodu sastavio Dr. Josif Pančić*]. Belgrade: Državna štamparija, 1865.

Pančić, Josif. *Flora of the Principality of Serbia or Vascular Plants that Grow Wild in Serbia. According to the Analytical Method Agreed by Dr. Josif Pančić* [*Flora Kneževine Srbije ili vaskularne biljke, koje u Srbiji divlje rastu. Po analitičkom metodu složio Dr. Josif Pančić*]. *Flora Principatus Serbiae*. Belgrade: Državna štamparija, 1874.

Pančić, Josif. *Additions to the Flora of the Principality of Serbia* [*Dodatak Flori kneževine Srbije*]. Beograd: Kraljevska srpska državna štamparija, 1884.

Pantocsek, Joseph. *Adnotationes ad Floram et Faunam Hercegovinae, Crnagorae et Dalmatiae*. Verh. des Ver. für Naturkunde in Presburg, *2*, 1874. [*Notes on the Flora and Fauna of Hercegovina, Montenegro, and Dalmatia*]

Parolly, Gerald. "Die Orchideen flora Montenegros." *AHO Baden-Wiirttemberg* 24 no. 2 (1992): 141–391.

Pedersen, Henrik Ærenlund, and Niels Faurholdt. *Ophrys: the Bee Orchids of Europe*. Kew: Royal Botanic Gardens, Kew Publishing, 2007.

Pedersen, Henrik Æ., Santi Watthana, and Kanok-Orn Srimuang. "Orchids in the Torrent: On the Circumscription, Conservation and Rheophytic Habit of *Epipactis flava*." *Botanical Journal of the Linnean Society* 172, no. 3 (2013): 358–70.

Petrović, Sava. *Flora of the Vicinity of Niš* [*Flora okoline Niša*]. Belgrade: Kraljevsko–srpska državna štamparija, 1882.

Petrović, Sava. *Dodatak flori okoline Niša* [*Addition to the Flora* in the *Environs* of *Niš*]. Belgrade: Kraljevsko–srpska državna štamparija, 1885.

Pillon, Yohan, Michael F. Fay, Alexey B. Shipunov, and Mark W. Chase. "Species Diversity versus Phylogenetic Diversity: A Practical Study in the Taxonomically Difficult Genus *Dactylorhiza* (Orchidaceae)." *Biological Conservation* 129, no. 1 (2006): 4–13.

Ponert, Jan. "Contribution to the Orchids of Republic of Macedonia and Serbia." *Journal Europäischer Orchideen* 46 (2014): 561–577.

Pridgeon, A. M., R. M. Bateman, A. V. Cox, J. R. Hapeman, and M. W. Chase. "Phylogenetics of Subtribe Orchidinae (Orchidoideae, Orchidaceae) Based on Nuclear ITS sequences. 1. Intergeneric Relationships and Polyphyly of *Orchis sensu lato*." *Lindleyana* 12 (1997): 89–109.

Pridgeon, M. Alec, Phillip J. Cribb, Mark W. Chase, and Finn N. Rasmussen, eds. *Genera Orchidacearum vol. 1. General introduction, Apostasioideae, Cypripedioideae*. New York: Oxford University Press Inc., 1999.

Pridgeon, M. Alec, Phillip J. Cribb, Mark W. Chase, and Finn N. Rasmussen, eds. *Genera Orchidacearum vol. 2. Orchidoideae (Part 1)*. New York: Oxford University Press Inc., 2001.

Pridgeon, M. Alec, Phillip J. Cribb, Mark W. Chase, and Finn N. Rasmussen, eds. *Genera Orchidacearum vol. 4. Epidendroideae (Part 1)*. New York: Oxford University Press Inc., 2005.

Prodán, Gyula. "Bács-Bodrog vármegye flórája." *Magyar Botanikai Lapok* 14, no. 5/12 (1916): 120–269.

Pulević, Vukić. *Građa za vaskularnu floru Crne Gore. Dopuna Conspectus Florae Montenegrinae J. Rohlene*. Podgorica: Republički zavod za zaštitu prirode Crne Gore, 2005. [*Material for the vascular flora of Montenegro. Supplement to Conspectus Florae Montenegrinae J. Rohlene]*

Radak, Boris. *Morphological Variability of Species of the Genus Anacamptis Rich. (Orchidoideae,* Orchidaceae*) in the Balkan Peninsula and the Pannonian Plain*. PhD diss., University of Novi Sad, 2019.

Radak, Boris Đ., Bojana Bokić S., Milica Rat M., Ranko Perić D., and Goran Anačkov, T. "Tribe Orchideae (Dressler & Dodson) Verm. 1977 (Orchidaceae, Orchidoideae) in Vojvodina." Paper presented at the *Botanical Symposium "Third Century of Botany in Vojvodina," Novi Sad, Serbia, April 15, 2016*. https://www.maticasrpska.org.rs/wordpress/assets/Knjiga-sazetaka.pdf.

Radak, B., J. Peškanov, A. Vlku, M. Prodanović, and G. Anačkov. "Orchids in Serbia: Additions to the An Annotated Checklist of Vascular Flora of Serbia." Paper presented at the *13th Symposium on the Flora of Southeastern Serbia and Neighboring Regions, Stara planina Mt., Serbia, June 20–23, 2019a*. http://www.sfses.com/docs/Book-of-Apstracts.pdf.

Radak, Boris, Dj, Aleksa Vlku Z, Jovan Peškanov M, Vlado Matevski S, and Goran Anačkov T. "Morphological Characterization of Three Natural Hybrid Orchid Taxa, New for Serbia, Montenegro and North

Macedonia." *Archives of Biological Sciences* 71, no. 4 (2019b): 596–607.

Rakosy, Demetra, Martin Streinzer, Hannes F. Paulus, and Johannes Spaethe. "Floral Visual Signal Increases Reproductive Success in a Sexually Deceptive Orchid." *Arthropod-Plant Interactions* 6, no. 4 (2012): 671–81.

Ramírez, Santiago R., Barbara Gravendeel, Rodrigo B. Singer, Charles R. Marshall, and Naomi E. Pierce. "Dating the Origin of the Orchidaceae from a Fossil Orchid with Its Pollinator." *Nature* 448, no. 7157 (2007): 1042–45.

Raunkiaer, Christen C. *The Life Forms of Plants and Statistical Plant Geography*. Oxford: Clarendon, 1934.

Rechinger, Karl H. "Ergebnisse einer botanischen Reise in den Bertiscus (Nordalbanische Alpen)." *Feddes Repertorium specierum novarum regni vegetabilis* 38, no. 13–25 (1935): 138–152; 319–389.

Rewicz, Agnieszka, Radomir Jaskuła, Tomasz Rewicz, and Grzegorz Tończyk. "Pollinator Diversity and Reproductive Success of *Epipactis helleborine* (L.) Crantz (Orchidaceae) in Anthropogenic and Natural Habitats." *PeerJ* 5 (2017): e3159.

Rumy, Károly. "Über das Klima Sirmiens." In *Die Versammlungen ungarischer Ärzte und Naturforscher: mit besonderer Beziehung auf die am 4. August 1842 zu Neusohl abgehaltene dritte Versammlung*, edited by Christian Andreas Zipser, 48–55. Neusohl [Banská Bystrica], 1846. ["About the climate of Sirmia." In *The Meetings of Hungarian Doctors and Naturalists: with special reference to the third meeting held on August 4, 1842 in Neusohl*]

Salzmann, C. C., and F. P. Schiestl. "Odour and Colour Polymorphism in the Food-Deceptive Orchid *Dactylorhiza romana*." *Plant Systematics and Evolution* 267, no. 1–4 (2007): 37–45.

Savić, Dragiša. *Ecology, Distribution and Conservation of the Species of the Family* Orchidaceae *on Mt. Fruška Gora.* Master's Thesis, University of Belgrade, 1998.

Savić, Dragiša, Goran Anačkov, and Pal Boža. "New Chorological Data for Flora of the Pannonian Region of Serbia." *Central European Journal of Biology* 3, no. 4 (2008): 461–70.

Schatz, Bertrand, Pierre Demougeot, and Gaëtan Gotanègre. "Insect Pollination and Inflorescence Structure in Two Species of *Spiranthes* Orchids." *In: Proceedings of the 4th International workshop on functional-structural plant models (FSPM), abstracts of papers and posters, 7–11 June 2004, Montpellier, France.* https://agritrop.cirad.fr/523905/.

Schiebold, Julienne M-I, Martin I. Bidartondo, Florian Lenhard, Andreas Makiola, and Gerhard Gebauer. "Exploiting Mycorrhizas in Broad Daylight: Partial Mycoheterotrophy Is a Common Nutritional Strategy in Meadow Orchids." *Journal of Ecology* 106, no. 1 (2018): 168–78.

Schiestl, F. P., M. Ayasse, H. F. Paulus, C. Löfstedt, B. S. Hansson, F. Ibarra, and W. Francke. "Sex Pheromone Mimicry in the Early Spider Orchid (*Ophrys sphegodes*): Patterns of Hydrocarbons as the Key Mechanism for Pollination by Sexual Deception." *Journal of Comparative Physiology. A, Sensory, Neural, and Behavioral Physiology* 186, no. 6 (2000): 567–74.

Schneller, August. "Beiträge zur Kenntniss der Phanerogamenflora von Futtak bei Peterwardein." *Verhandlungen der Kaiserlich-Königlichen Zoologisch-Botanischen Gesellschaft in Wien* 3 (1858): 1–22. ["Contributions to the knowledge of the phanerogam flora of Futtak near Peterwardein." *Negotiations of the Imperial-Royal Zoological-Botanical Society in Vienna*]

Schulzer, Stefan, August Kanitz, and Josef A. Knapp. "Die bisher bekannten Pflanzen Slavoniens." *Verhandlungen der Kaiserlich-Königlichen Zoologisch-Botanischen Gesellschaft in Wien* 16 (1866): 1–172. ["The previously known plants of Slavonia." *Negotiations of the Imperial-Royal Zoological-Botanical Society in Vienna*]

Scopece, Giovanni, Nicolas Juillet, Andreas Müller, Florian P. Schiestl, and Salvatore Cozzolino. "Pollinator Attraction in *Anacamptis papilionacea* (Orchidaceae): A Food or a Sex Promise?" *Plant Species Biology* 24, no. 2 (2009): 109–14.

Sedivy, Claudio, Christophe J. Praz, Andreas Müller, Alex Widmer, and Silvia Dorn. "Patterns of Host-Plant Choice in Bees of the Genus *Chelostoma*: The Constraint Hypothesis of Host-Range Evolution in Bees." *Evolution* 62, no. 10 (2008): 2487–2507.

Selosse, M. A., A. Faccio, G. Scappaticci, and P. Bonfante. "Chlorophyllous and Achlorophyllous Specimens of *Epipactis microphylla* (Neottieae, Orchidaceae) Are Associated with Ectomycorrhizal Septomycetes, Including Truffles." *Microbial Ecology* 47, no. 4 (2004): 416–26.

Selosse, Marc-André, Michael WEIss, Jean-Luc Jany, and Annie Tillier. "Communities and Populations of Sebacinoid Basidiomycetes Associated with the Achlorophyllous Orchid *Neottia nidus-avis* (L.) L.C.M. Rich. and Neighbouring Tree Ectomycorrhizae." *Molecular Ecology* 11, no. 9 (2002): 1831–44.

Silvera, Katia, Louis S. Santiago, John C. Cushman, and Klaus Winter. "Crassulacean Acid Metabolism and Epiphytism Linked to Adaptive Radiations in the Orchidaceae." *Plant Physiology* 149, no. 4 (2009): 1838–47.

Soó, R. "Revison der Orchideen Südosteuropas und Südwestasiens." *Botanisches Archiv* 23 (1929): 1–196.

Ståhlberg, David. "Habitat Differentiation, Hybridization and Gene Flow Patterns in Mixed Populations of Diploid and Autotetraploid *Dactylorhiza maculata* s.l. (Orchidaceae)." *Evolutionary Ecology* 23, no. 2 (2009): 295–328.

Ståhlberg, David, and Mikael Hedrén. "Systematics and Phylogeography of the *Dactylorhiza maculata* Complex (Orchidaceae) in Scandinavia: Insights from Cytological, Morphological and Molecular Data." *Plant Systematics and Evolution* 273, no. 1–2 (2008): 107–32.

Steiner, Kim E., V. B. Whitehead, and S. D. Johnson. "Floral and Pollinator Divergence in Two Sexually Deceptive South African Orchids." *American Journal of Botany* 81, no. 2 (1994): 185–94.

Stevanović, Vladimir. "Biogeographical division of the territory of Yugoslavia." In *Biodiversity of Yugoslavia with an Overview of Species of International Importance*, edited by Vladimir Stevanović, and Voislav Vasić, 117–127. Belgrade: Ecolibri and Faculty of Biology, 1995.

Stevanović, Vladimir (1992). "Floristic Division of the Territory of Serbia with an Overview of Higher Chorion and Appropriate Floral Elements." In *The Flora of Serbia I* edited by Miloje R. Sarić, 49–65. Belgrade: Serbian Academy of Sciences and Arts, 1992.

Stevanović, Vladimir, Slobodan Jovanović, Dmitar Lakušić, and Marjan Niketić. "Diversity of the Vascular Flora of Yugoslavia With an Overview of Species of International Importance." In *Biodiversity of Yugoslavia with an Overview of Species of International Importance*, edited by Vladimir Stevanović, and Voislav Vasić, 183–217. Belgrade: Ecolibri and Faculty of Biology, 1995.

Stevanović, Vladimir, and Branka Stevanović. "Basic Climatic, Geological and Pedological Factors of Biodiversity of Terrestrial Ecosystems of Yugoslavia." In *Biodiversity of Yugoslavia with an Overview of Species of International Importance*, edited by Vladimir Stevanović, and Voislav Vasić, 75–95. Belgrade: Ecolibri and Faculty of Biology, 1995.

Stevanović, V., M. Niketić, and D. Lakušić. "Chorological Additions to the Flora of Eastern Yugoslavia." *Flora Mediterranea* 1 (1991): 121–42.

Stevanović, Vladimir, Snežana Vukojičić, Jasmina Šinžar-Sekulić, Maja Lazarević, Gordana Tomović, and Kit Tan. "Distribution and Diversity of Arctic-Alpine Species in the Balkans." *Plant Systematics and Evolution* 283, no. 3–4 (2009): 219–35.

Stpiczyńska, Małgorzata, and Janusz Matusiewicz. "Anatomy and Ultrastructure of Spur Nectary of *Gymnadenia conopsea* (L.) Orchidaceae." *Acta Societatis Botanicorum Poloniae* 70, no. 4 (2001): 267–72.

Sun, Mimi, Karin Gross, and Florian P. Schiestl. "Floral Adaptation to Local Pollinator Guilds in a Terrestrial Orchid." *Annals of Botany* 113, no. 2 (2014): 289–300.

Süveges, Kristóf, Viktor Löki, Ádám Lovas-Kiss, Tibor Ljubka, Réka Fekete, Attila Takács, Orsolya Vincze, Balázs András Lukács, and Attila Molnár V. "From European Priority Species to Characteristic Apophyte: *Epipactis tallosii* (Orchidaceae)." *Willdenowia* 49, no. 3 (2019): 401–9.

Svolynskiy A. D., S. P. Ivanov, and A. V. Fateryga. "Peculiarities of Anthecology of the Provence Orchid (*Orchis provincialis*, Orchidaceae) in the Crimea: Pollinators, Method of Their Attraction, Pollination Rate." *Optimization and Protection of Ecosystems. Simferopol: TNU* 11 (2014): 144–157.

Swarts, Nigel D., and Kingsley W. Dixon. "Terrestrial Orchid Conservation in the Age of Extinction." *Annals of Botany* 104, no. 3 (2009): 543–56.

Šabanović, Elvedin, Aldin Boškailo, Vladimir Ranđelović, Samir Đug, Šemso Šarić, Safija Boškailo, and Sanida Bektić. *Orhideje Zeničko-dobojskog kantona.* Visoko: Zavičajni muzej, 2020. [*Orchids of Zenica-Doboj Canton.*]

Šoljan, Dubravka, Edina Muratović, and Sabaheta Abadžić. *Orhideje planina oko Sarajeva.* Sarajevo: Dobra knjiga, 2014. [*Orchids of the mountains around Sarajevo.*]

Štípková, Zuzana, Spyros Tsiftsis, and Pavel Kindlmann. "Distribution of Orchids with Different Rooting Systems in the Czech Republic." *Plants* 10, no. 4 (2021): 632.

Štípková, Zuzana, Spyros Tsiftsis, and Pavel Kindlmann, "Pollination Mechanisms Are Driving Orchid Distribution in Space," *Scientific Reports* 10, no. 1 (2020): 850.

Tali, Kadri, Michael F. Fay, and Richard M. Bateman. "Little Genetic Differentiation across Europe between Early-Flowering and Late-Flowering Populations of the Rapidly Declining Orchid *Neotinea ustulata.*" *Biological Journal of the Linnean Society* 87, no. 1 (2006): 13–25.

Tali, Kadri, Michael J. Y. Foley, and Tiiu Kull. "Biological flora of the British Isles, 232. *Orchis ustulata* L." *Journal of Ecology* 92, no. 1 (2004): 174–84.

Tatarenko, Irina. "Growth Habits of Temperate Terrestrial Orchids." In *Orchid Biology – Reviews and Perspectives* Vol. *IX*, edited by Kenneth M. Cameron, Joseph Arditti, and Tiiu Kull, 91–161. Bronx, New York: The New York Botanical Garden Press, 2007.

Taylor, Amanda, Gunnar Keppel, Patrick Weigelt, Gerhard Zotz, and Holger Kreft. "Functional Traits Are Key to Understanding Orchid Diversity on Islands." *Ecography* 44 (2021): 1–12, https://doi.org/10.1111/ecog.05410.

Teppner, Herwig, and Erich Klein. "Etiam atque etiam – *Nigritella* versus *Gymnadenia*: Neukombinationen und *Gymnadenia dolomitensis* spec. nova (Orchidaceae*: Orchideae*)." *Phyton (Horn)* 38 (1998): 220–224.

Taylor, Lin, and David L. Roberts. "Biological flora of the British Isles: *Epipogium aphyllum* Sw." *Journal of Ecology* 99, no. 3 (2011): 878–90.

Tichy, H., and C. Del Prete. "Chromosome Numbers of European Orchids: Interpreting Data." *Journal Europäischer Orchideen* 33 (2001): 217–234.

Tomović, Gordana, Marjan Niketić, Dmitar Lakušić, Vladimir Ranđelović, and Vladimir Stevanović. "Balkan Endemic Plants in Central Serbia and Kosovo Regions: Distribution Patterns, Ecological Characteristics, and Centres of Diversity: Balkan Endemic Plants in Central Serbia and Kosovo Regions." *Botanical Journal of the Linnean Society* 176 (2014): 173–202.

Tomović, Gordana, Marko Sabovljević S, Ivana Djokić, Predrag Petrović, Vladan Djordjević, Predrag Lazarević, Ermin Mašić, et al. "New Records and Noteworthy Data of Plants, Algae and Fungi in SE Europe and Adjacent Regions, 2." *Botanica Serbica* 44, no. 2 (2020): 251–59.

Tomović, Gordana, Snezana Vukojićić, M. Niketić, and D. Lakušić. "New Chorological Data on Some Threatened and Rare Plants in Serbia." *Archives of Biological Sciences* 59, no. 1 (2007): 63–73.

Tranchida-Lombardo, Valentina, Donata Cafasso, Antonia Cristaudo, and Salvatore Cozzolino. "Phylogeographic Patterns, Genetic Affinities and Morphological Differentiation between *Epipactis helleborine* and Related Lineages in a Mediterranean Glacial Refugium." *Annals of Botany* 107, no. 3 (2011): 427–36.

Tremblay, Raymond L., James D. Ackerman, Jess K. Zimmerman, and Ricardo N. Calvo. "Variation in Sexual Reproduction in Orchids and Its Evolutionary Consequences: A Spasmodic Journey to Diversification." *Biological Journal of the Linnean Society* 84, no. 1 (2005): 1–54.

Tsiftsis, Spyros. "Morphological Variability of *Himantoglossum* s.s. (Orchidaceae) in Greece." *Phytotaxa* 245, no. 1 (2016): 17–30.

Tsiftsis, Spyros. "The Complex Effect of Heterogeneity and Isolation in Determining Alpha and Beta Orchid Diversity on Islands in the Aegean Archipelago," *Systematics and Biodiversity* 18, no. 3 (2020): 281–94.

Tsiftsis, Spyros, and Zissis Antonopoulos. *Atlas of the Greek Orchids, Vol I*. Rethymno, Greece: Mediterraneo Editions, 2017.

Tsiftsis, Spyros, Vladan Djordjević, and Ioannis Tsiripidis. "*Neottia cordata* (Orchidaceae) at Its Southernmost Distribution Border in Europe: Threat Status and Effectiveness of Natura 2000 Network for Its Conservation." *Journal for Nature Conservation* 48 (2019a): 27–35.

Tsiftsis, Spyros, and Vladan Djordjević. "Habitat Effects and Differences in the Reproductive Success of *Orchis punctulata* and *Orchis purpurea* (Orchidaceae)." *Turkish Journal of Botany* 42, no. 4 (2018): 400–411.

Tsiftsis, Spyros, and Vladan Djordjević, "Modelling Sexually Deceptive Orchid Species Distributions under Future Climates: The Importance of Plant-Pollinator Interactions." *Scientific Reports* 10, no. 1 (2020): 10623.

Tsiftsis, Spyros, and Ioannis Tsiripidis. "Temporal and Spatial Patterns of Orchid Species Distribution in Greece: Implications for Conservation." *Biodiversity and Conservation* 29, no. 11 (2020): 3461–89.

Tsiftsis, Spyros, Zuzana Štípková, and Pavel Kindlmann. "Role of Way of Life, Latitude, Elevation and Climate on the Richness and Distribution of Orchid Species." *Biodiversity and Conservation* 28, no. 1 (2019b): 75–96.

Tyteca, Daniel, Manon Ceinos, Jean-Louis Gathoye, Rein Brys, and Hans Jacquemyn. "On the Morphological, Biological and Genetic Heterogeneity of the Genus *Orchis* (Orchidaceae, *Orchidinae*)." *Phytotaxa* 75, no. 1 (2012): 19–32.

Vakhrameeva, M. G., I. V. Tatarenko, T. I. Varlygina, G. K. Torosyan, and M. N. Zagulskii. *Orchids of Russia and Adjacent Countries (within the borders of the former USSR).* Ruggell/Liechtenstein: A.R.G. Gantner Verlag, 2008.

Vallius, E. "Position-Dependent Reproductive Success of Flowers in *Dactylorhiza maculata* (Orchidaceae): Position-Dependent Reproductive Success of Flowers." *Functional Ecology* 14, no. 5 (2000): 573–79.

Vallius, E., V. Salonen, and T. Kull. "Factors of Divergence in Co-Occurring Varieties of *Dactylorhiza incarnata* (Orchidaceae)." *Plant Systematics and Evolution* 248, no. 1–4 (2004): 177–189.

van der Cingel, Nelis A. *An Atlas of Orchid Pollination. European orchids*. Rotterdam: A.A. Balkema, 1995.

van der Pijl, L., and Calaway H. Dodson. *Orchid Flowers: Their Pollination and Evolution*. Coral Gables, FL, USA: University of Miami Press, 1966.

Vangjeli, J., B. Ruci, A. Mullaj, K. Paparisto, and Xh. Qosja. *Flora e Shqipërisë Vol. 4.* [*Flora of Albania 4*]. Tiranë: Akademia e Shkencave e Republikes se Shqiperise, 2000.

Vereecken, Nicolas J., Amots Dafni, and Salvatore Cozzolino. "Pollination Syndromes in Mediterranean Orchids – Implications for

Speciation, Taxonomy and Conservation." *Botanical Review* 76, no. 2 (2010): 220–40.

Vogel, Stefan. *Pollination von Orchis papilionacea L. in den Schwarmbahnen von Eucera tuberculata F. Jahresberichte des Naturwissenschaftlichen Vereins Wuppertal* 25 (1972): 67–74. [*Pollination of Orchis papilionacea L. in the swarm tracks of Eucera tuberculata F. Annual reports of the Natural Science Association Wuppertal*]

Vogt-Schilb, Hélène, François Munoz, Franck Richard, and Bertrand Schatz. "Recent Declines and Range Changes of Orchids in Western Europe (France, Belgium and Luxembourg)." *Biological Conservation* 190 (2015): 133–41.

Vogt-Schilb, Hélène, Roger Pradel, Philippe Geniez, Laetitia Hugot, Alain Delage, Franck Richard, and Bertrand Schatz. "Responses of Orchids to Habitat Change in Corsica over 27 Years." *Annals of Botany* 118, no. 1 (2016): 115–23.

Voth, Walter. "Bestäubungsbiologische Beobachtungen an *Orchis militaris* L." *Die Orchidee* 38, no. 2 (1987): 77–84. [Pollination biological observations on *Orchis militaris* L. *The orchid*]

Vöth, Walter. "Die 'ausgeborgten' Bestäuber von *Orchis pallens* L." *Orchidee* 33 no. 5 (1982): 196–203. [The 'borrowed' pollinators of *Orchis pallens* L. *Orchid*]

Vöth, Walter. "*Echinomyia magnicornis* Zett. Bestäuber von *Orchis ustulata* L." *Die Orchidee* 35 (1984): 189–192. ["Echinomyia magnicornis Zett. Bestäuber von *Orchis ustulata* L. " *The orchid*]

Vöth, Walter. "*Gymnadenia nigritella* und ihre Bestäuber." *Journal Europäischer Orchideen* 32, no. 3–4 (2000): 547–573. [*Gymnadenia nigritella* and its pollinators. *Journal of European Orchids*]

Vöth, Walter. "Lebensgeschichte und Bestäuber der Orchideen am Beispiel von Niederösterreich." *Stapfia* 65 (1999): 182–196. [Life story and pollinators of orchids using the example of Lower Austria.]

Vöth, Walter. "*Trielis villosa* var. *rubra*, Bestäuber von *Orchis coriophora*." *Die Orchidee* 26 (1975): 170–172. [*Trielis villosa* var. rubra, pollinator of *Orchis coriophora. The orchid*]

Vukojičić, Snežana, Ksenija Jakovljević, Vlado Matevski, Vladimir Randjelović, Marjan Niketić, and Dmitar Lakušić. "Distribution, Diversity and Conservation of Boreo-Montane Plant Species in the Central Part of the Balkan Peninsula and the Southern Part of the Pannonian Plain." *Folia Geobotanica* 49, no. 4 (2014): 487–505.

Waterman, Richard J., and Martin I. Bidartondo. "Deception above, Deception below: Linking Pollination and Mycorrhizal Biology of Orchids." *Journal of Experimental Botany* 59, no. 5 (2008): 1085–96.

WCSP. "World Checklist of Selected Plant Families." *Kew: Royal Botanic Gardens*, http://apps.kew.org/wcsp/, 2020.

Wettstein, Richard. "Beitrag zur Flora Albaniens. Bearbeitung der von J. Dörfler im Jahre 1890 im Gebiete des Šar-Dagh gesammelten Pflanzen." *Bibliotheca Botanica* 5, no. 26 (1892): 1–103.

Whigham, Dennis F., and Willems, J. H. 2003. "Demographic Studies and Life-History Strategies of Temperate Terrestrial Orchids as a Basis for Conservation." in *Orchid Conservation*, edited by Kingsley, Dixon W., Shelagh Kell P., R. L. Barrett, and Philip Cribb, 137–158. Kota Kinabaluk Sabah, Malaysia: Natural History Publications, 2003.

Wood, Jeffrey, and Margaret Ramsay. "Plate 482. *Anacamptis laxiflora* Orchidaceae." *Curtis's botanical Magazine* 21, no. 1 (2004): 26–33.

Wraith, Jenna, and Catherine Pickering. "Quantifying Anthropogenic Threats to Orchids Using the IUCN Red List." *Ambio* 47, no. 3 (2018): 307–17.

Wróblewska, Ada, Lech Szczepaniak, Andrzej Bajguz, Iwona Jędrzejczyk, Izabela Tałałaj, Beata Ostrowiecka, Emilia Brzosko, Edyta Jermakowicz, and Paweł Mirski. "Deceptive Strategy in *Dactylorhiza* Orchids: Multidirectional Evolution of Floral Chemistry." *Annals of Botany* 123, no. 6 (2019): 1005–16.

Zorkóczy, L. *Újvidék és környékének flórája: Üjvidék* [Novi Sad]: Popovits M. Testvérek Könyvnyomdája, 1896. [*Flora of Novi Sad and its surroundings*: Novi Sad: Popovits M. Brothers Book Printing House, 1896.]

Zotz, Gerhard, and Uwe Winkler. "Aerial Roots of Epiphytic Orchids: The Velamen Radicum and Its Role in Water and Nutrient Uptake." *Oecologia* 171, no. 3 (2013): 733–41.

Reviewed by:

Spyros Tsiftsis

Department of Forest and Natural Environment Sciences, International Hellenic University, 1st km Dramas – Mikrochoriou, Drama, Greece

Agnieszka Rewicz

Department of Biogeography, Paleoecology and Nature Conservation, Faculty of Biology and Environmental Protection, University of Lodz, Lodz, Poland

In: Orchidaceae
Editor: Vladan Djordjević

ISBN: 978-1-68507-189-9

Chapter 4

ORCHIDACEAE IN THE REPUBLIC OF MORDOVIA (RUSSIA): DISTRIBUTION AND CONSERVATION STATUS

Anatoliy A. Khapugin[1,2,*] and Gennadiy G. Chugunov[3]
[1]Institute of Environmental and Agricultural Biology (X-BIO), Tyumen State University, Tyumen, Tyumen region, Russia
[2]Joint Directorate of the Mordovia State Nature Reserve and National Park "Smolny", Saransk, Republic of Mordovia, Russia
[3]Mordovia State University, Saransk, Republic of Mordovia, Russia

ABSTRACT

The Republic of Mordovia is situated in the center of European Russia. It belongs to the zones of coniferous and deciduous forests and forest-steppes. Here are the geographical limits of distribution of many vascular plant species. Orchids are among the most vulnerable plants in many regions of the world.

[*] Corresponding Author's E-mail: hapugin88@yandex.ru.

They are threatened with extinction in a high diversity of habitats, especially since most natural ecosystems in the study region are under anthropogenic pressure. For example, most steppe sites are used as arable lands and pasture, while forests are affected by deforestation and forest fires. The most intact and least disturbed ecosystems are found on sites unsuitable for agriculture and in protected areas. Twenty-four species of orchids are known to occur in the Republic of Mordovia. Many of them are included in the Red Data Book of the Republic of Mordovia and in the Red Data Book of the Russian Federation. The network of protected areas contributes to the conservation of both natural ecosystems and orchid populations, including threatened species. Population-based studies concern some of the orchid species in the Republic of Mordovia. These data provide valuable insights into the conservation status and habitat preferences of orchids. In this chapter, we analyze the distribution and conservation issues of the orchid flora in the Republic of Mordovia, based on systematized and updated data on their distribution and taxonomy. The hotspots of orchid diversity in the study area are indicated, and the problems in orchid research and possible ways to solve them are highlighted.

Keywords: orchid species, conservation, red data book, Republic of Mordovia, protected area

INTRODUCTION

The Orchidaceae are considered one of the richest plant families in the world, comprising more than 28 000 accepted species and 763 genera (Christenhusz and Byng 2016). Orchids occur in a variety of habitats, but most commonly in forests (e.g., Khapugin 2020). They are presented by epiphytic, terrestrial, and subterranean life forms (e.g., Cribb et al. 2003; Givnish et al. 2016). Many orchids are endemics, naturally rare and threatened taxa (e.g., Barberena et al. 2018; Fateryga et al. 2020; Besi et al. 2020). More than half (56.5%) of the 948 orchid species assessed worldwide are considered threatened according to the Global IUCN Red List (e.g., Gale et al. 2018; Kottawa-Arachchi and Gunasekara 2020). Given the ever-increasing extinction rates of plant species worldwide, more action is needed (e.g., Le Roux et al. 2019).

In addition to the global IUCN assessments (e.g., endemics, see Wagensommer et al. 2020), there are numerous publications on the regional assessment of orchid species using the IUCN Red List Categories and Criteria (e.g., Bernardos et al. 2006; Eliáš et al. 2015; Kull et al. 2016; Djordjević et al. 2017; Khapugin et al. 2017a,c; de la Torre Llorente 2018; Juiling et al. 2020). Orchidaceae taxa are continuously described annually, mainly in the tropics and subtropics (e.g., Ječmenica et al. 2016). Specifically, the orchid richness and abundance depend on weather conditions (e.g., Djordjević and Tsiftsis 2020; Kirillova and Kirillov 2020; Mursal et al. 2020; Romano et al. 2020), habitat degradation (e.g., Tsiftsis et al. 2008; Djordjević and Tsiftsis 2020; Tatarenko et al. 2020), altitude (e.g., Tsiftsis et al. 2008; Acharya et al. 2011; Crain and Tremblay 2014; Djordjević et al. 2016a,b; Djordjević and Tsiftsis 2020), light availability (e.g., Djordjević et al. 2016b; Perazza and Decarli 2020), soil moisture (e.g., Djordjević et al. 2016a,b; Mursal et al. 2020), and canopy height and area (e.g., Gravendeel et al. 2004; Huang et al. 2008; Tsiftsis et al. 2008; Djordjević et al. 2016b). Often, data on population status, distribution, or conservation importance are obtained primarily through molecular phylogenetic methods (e.g., Terentieva et al. 2020), studies of morphological variation (e.g., Shirokov et al. 2020), or laboratory-based studies of orchids and their environmental preferences (e.g., Calevo and Bazzicalupo 2020). Therefore, studies on all aspects of orchid biology and ecology are important to gain essential knowledge on the diversity and conservation of the Orchidaceae family.

Orchids in Russia are represented by 135 species and 13 subspecies belonging to 38 genera (Efimov 2020), with the highest taxonomic diversity in the Russian Far East and South Russia (e.g., Popovich et al. 2020). Twenty-four species of orchids are known in the Republic of Mordovia, located in the center of European Russia (Silaeva et al. 2010). Available data on distribution (Silaeva et al. 2010; Khapugin and Silaeva 2020), population status (e.g., Khapugin et al. 2016a,b, 2017c; Khapugin and Korochkina 2017) and conservation issues (Khapugin et al. 2017a,b; Silaeva 2017) of Mordovian orchids are found in several publications.

Therefore, there is no single source where all these data are available together. So, the present chapter aims to generalize and analyze the data on orchid species known for the Republic of Mordovia.

METHODS

Study Area

The study was conducted in the Republic of Mordovia (hereafter Mordovia), with an area of 26 200 km^2 (Figure 1), located in the center of European Russia, on the border of the forest and forest-steppe zones, resulting in a high habitat diversity. Coniferous and mixed forests (i.e., coniferous and deciduous) are distributed in the western and northwestern parts of the region. Broad leaved forests are located in the central and eastern parts of Mordovia. Forest-steppe landscapes dominate eastern and southeastern Mordovia (Yamashkin 2012).

Distribution of Orchids

A grid mapping scheme proposed by Grishutkin (2013) was applied to show the current distribution of orchids and hotspots/coldspots of orchid diversity in Mordovia. This scheme has 309 grid cells, each with an area of about 100 km^2 (10 × 10 km). Data on the distribution of threatened species were taken from the Red Data Book of the Republic of Mordovia (Silaeva 2017) and recently published materials (Khapugin et al. 2017d; Ageeva et al. 2020). Data on common species were obtained from the herbarium collections of GMU and HMNR. Although Silaeva et al. (2010) indicated a higher number of orchid sites in the region, there was neither herbarium evidence nor any publication confirming such statements. Finally, orchid distribution data were obtained from the online iNaturalist database (Research Grade status only).

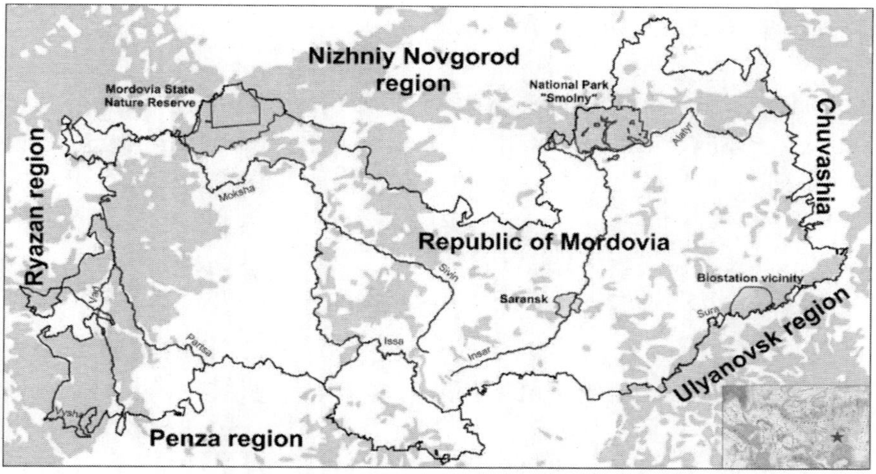

Figure 1. Biodiversity research centers in the Republic of Mordovia [Khapugin and Silaeva (2020) with modificationsn]. The inset in the right corner shows the location of the study area in Europe.

Taxonomical Issues of the Orchid Flora in the Republic of Mordovia

Optimal collaboration between taxonomists and ecologists is a hot topic worldwide (Halme et al. 2015), since a relatively small number of ecological publications refer to modern taxonomy and nomenclature of plants using international databases, such as POWO (2020) or The PlantList (2013) (Khapugin, unpublished). Accordingly, most summaries of the regional flora in Russia cite irrelevant (i.e., synonymous) species names of plants, including those referring to the flora of Mordovia. In this chapter, we focus on orchid species, according to modern taxonomic reviews, following POWO (2020), currently having a different accepted name than the one cited in the regional synopsis of the flora (Silaeva et al. 2010) and the Red Data Book of Mordovia (Silaeva 2017).

Data Processing and Software

Mapping and measurements were processed using MapInfo Pro software. Statistical calculations were performed using PAST 3.26 software (Hammer et al. 2001), Microsoft Excel package.

RESULTS AND DISCUSSION

Flora and Distribution of Orchids in the Republic of Mordovia

The locations of orchid species in Mordovia cover 133 of 309 cells (43.0%) of the regional grid map (Figure 2, 3, 4). Certain orchid species are characterized by different numbers of sites and distribution patterns in the study region. For example, *Cypripedium guttatum* L. (Figure 2D), *Dactylorhiza viridis* (L.) R.M. Bateman, Pridgeon & M.W. Chase (Figure 2H), *Hammarbya paludosa* (L.) Kuntze (Figure 3B), *Neotinea ustulata* (L.) R.M. Bateman, Pridgeon & M.W. Chase (Figure 3E) occur in one grid cell each, while *Neottia cordata* (L.) Rich. occurs in two grid cells (Figure 3F). Several orchid species, on the other hand, are represented by a relatively high number of sites, namely *Platanthera bifolia* (L.) Rich. (56 grid cells, Figure 3J), *Epipactis helleborine* (L.) Crantz (53 grid cells, Figure 2I), *Dactylorhiza fuchsii* (Druce) Soó (32 grid cells, Figure 2E), *D. incarnata* (L.) Soó (31 grid cells, Figure 2F), and *Neottia nidus-avis* (L.) Rich. (30 grid cells, Figure 3G). These species are the most common orchids in Mordovia, although they should be given special attention being highly vulnerable.

The grid map visualizing orchid distribution shows the pattern of total species number without a number of records relating to the same sites over the period of the botanical surveys in Mordovia (Figure 4).

However, we believe that the distribution pattern of the most common species (see above) is far from complete, as many floristic records are still not published or supported by herbarium vouchers or kept in researchers' personal notebooks. Therefore, it is almost impossible to to give a complete overview. Hence, there is an urgent need to either publish these records or confirm them with herbarium vouchers or photo-observations uploaded to the iNaturalist portal or in some other way that allows confirming or denying the presence of a particular orchid species in the region.

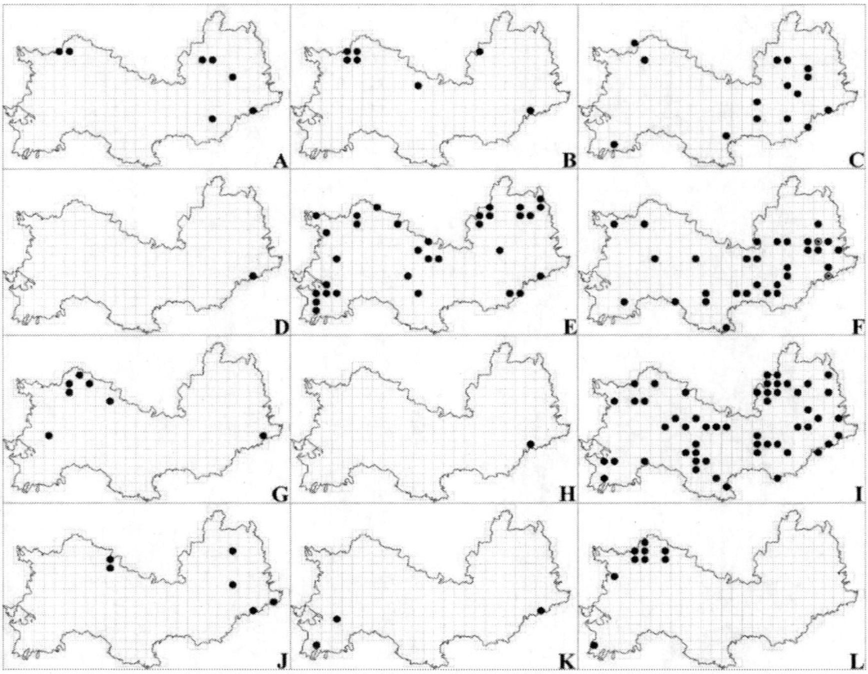

Figure 2. The distribution of orchid species in the Republic of Mordovia. Designations: A: *Cephalanthera rubra*; B: *Corallorhiza trifida*; C: *Cypripedium calceolus*; D: *Cypripedium guttatum*; E: *Dactylorhiza fuchsii*; F: *Dactylorhiza incarnata* (special signs indicates grid cells, where *Dactylorhiza incarnata* subsp. *cruenta* is known); G: *Dactylorhiza maculata*; H: *Dactylorhiza viridis*; I: *Epipactis helleborine*; J: *Epipactis palustris*; K: *Epipogium aphyllum*; L: *Goodyera repens.*

Finally, we counted the number of orchid species per grid cell in Mordovia, confirming previous assumptions about an orchid hotspot near the Mordovian State University Biostation (East Mordovia). The grid cells assigned to this proposed protected area contain 15 orchid species (Figure 4). Another area in northwestern Mordovia contains 13 orchid species, adjacent to grid cells with eight and nine species. These areas coincide with the Mordovia State Nature Reserve, a federal protected area and the richest biodiversity research center in the Republic of Mordovia (Khapugin and Silaeva 2020).

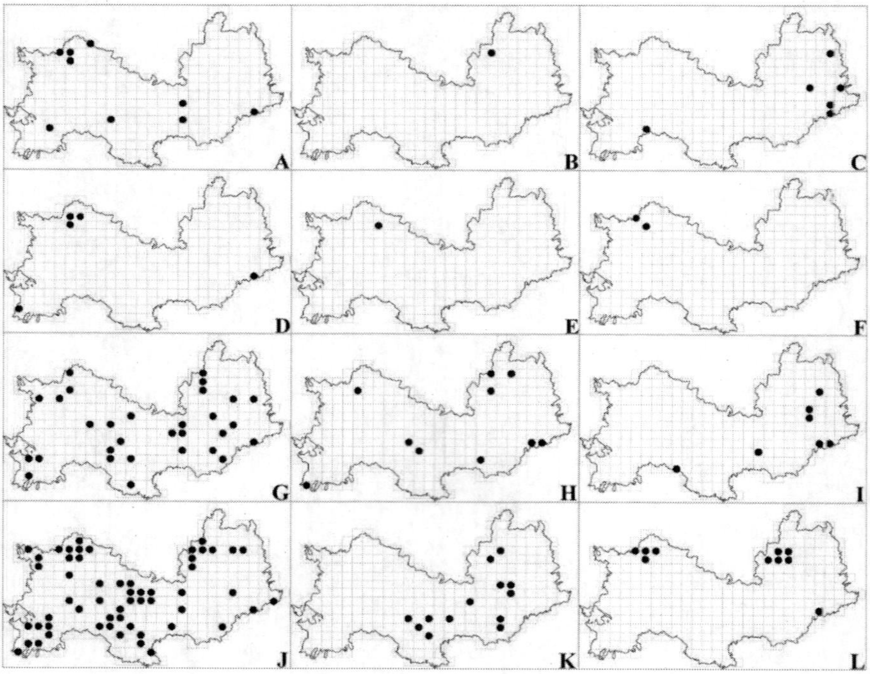

Figure 3. The distribution of orchid species in the Republic of Mordovia. Designations: A: *Gymnadenia conopsea*; B: *Hammarbya paludosa*; C: *Herminium monorchis*; D: *Malaxis monophyllos*; E: *Neotinea ustulata*; F: *Neottia cordata*; G: *Neottia nidus-avis*; H: *Neottia ovata*; I: *Orchis militaris*; J: *Platanthera bifolia*; K: *Platanthera chlorantha*; L: *Ponerorchis cucullata*.

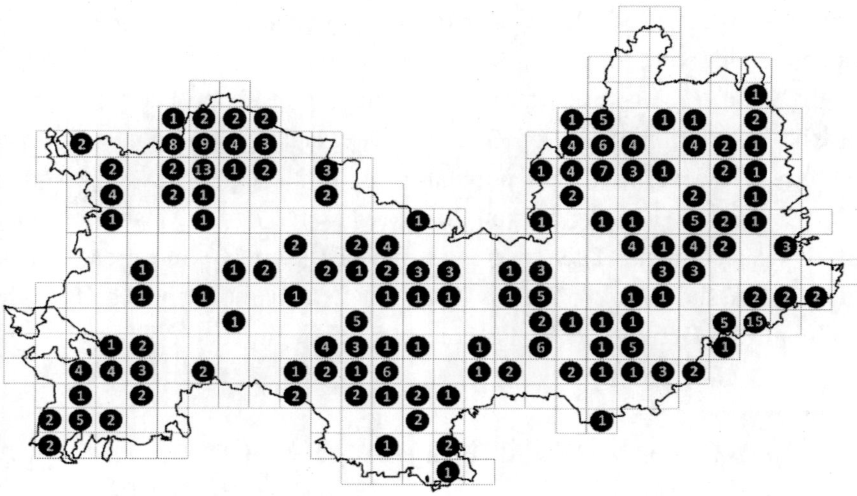

Figure 4. Orchid species diversity per grid cell in the Republic of Mordovia (Russia).

Orchid Flora of the Republic of Mordovia

Cephalanthera rubra (L.) Rich.

Cephalanthera rubra (L.) Rich. occurs in 7 grid cells (2.3% of the total of 309 grid cells) (Figure 2A). The species is legally protected in the region by being included in the Red Data Book of the Republic of Mordovia (rarity category 2, vulnerable species) (Silaeva 2017) and has recently been classified as a Critically Endangered taxon, according to the IUCN Red List categories (Khapugin et al. 2017a). The population density is low at all regional sites. Population-based studies of this orchid have not been conducted in Mordovia. Targeted attempts to recover its populations in Mordovia State Nature Reserve are still unsuccessful (Khapugin et al. 2012).

Corallorhiza trifida Châtel.

Corallorhiza trifida Châtel. locations occur in 7 grid cells (2.3%) (Figure 2B). This species is legally protected. It is included in the Red Data Book of the Republic of Mordovia (rarity category 1, endangered species) (Silaeva 2017) and has recently been assessed as a Critically Endangered taxon, according to the IUCN Red List categories (Khapugin et al. 2017a). The overall abundance of this orchid is probably low. Data on the population status of *C. trifida* are lacking, due to its specific biology. However, a recent population of this species was surveyed in 2014 and 2015 (Khapugin et al. 2015), where *C. trifida* was represented by 3–43 specimens. This species is one of the most widespread orchid taxa in Russia (Efimov 2020). It is included in more than half of the regional Red Data Books in this country (Khapugin et al. 2020).

Cypripedium calceolus L.

Cypripedium calceolus L. (Figure 5) occurs in 15 grid cells (4.9%) (Figure 2C). This species is legally protected in the Republic of Mordovia. It is included in the Red Data Book of Mordovia (rarity category 2, vulnerable species) (Silaeva 2017). *Cypripedium calceolus* has recently been assessed as an Endangered taxon, according to the IUCN Red List categories (Khapugin et al. 2017a). This species is one of the best known orchid species in Mordovia. Four of the known *C. calceolus* populations in Mordovia have been studied for more than ten years (Khapugin et al. 2014, 2017c), previously by the staff of the Mordovia State Nature Reserve (Sanaeva 2011). The most viable population of this orchid is located in the Mordovia State Nature Reserve (Khapugin et al. 2014, 2017c). In Russia, this species is one of the most widespread orchid taxa (Efimov 2020), most frequently included in the regional Red Data Books (Khapugin et al. 2020). Monitoring of *C. calceolus* should be continued in Mordovia, which will allow comparison of regional data with similar records obtained in the species' range (Egorova et al. 2020; Perazza and Decarli 2020).

Cypripedium guttatum Sw.

Cypripedium guttatum Sw. (Figure 6) occurs in one grid cell (0.3%) (Figure 2D). This species is legally protected being included in the Red Data Book of the Republic of Mordovia (rarity category 1, endangered species) (Silaeva 2017). Recently, it has been classified as a Critically

Endangered taxon, according to the IUCN Red List categories (Khapugin et al. 2017a). Targeted population-based studies have not been conducted in Mordovia yet, and the only known population is monitored annually to date. In Russia, this species is considered one of the widely distributed orchid taxa (Efimov 2020). It is included in the Red Data Books of more than half of all the Russian regions (Khapugin et al. 2020). We strongly recommend starting annual population-based studies of *C. guttatum*, taking into account that it could concern one of the *C. guttatum* forms mentioned by Shirokov et al. (2020).

Dactylorhiza fuchsii (Druce) Soó

Dactylorhiza fuchsii (Druce) Soó sites occur in 32 grid cells (10.4%) (Figure 2E). In Mordovia, this orchid species inhabits moist forests, forest glades, and swamp meadows (Silaeva et al. 2010). It is listed in the additional list of Red Data Book of the Republic of Mordovia (Silaeva 2017) as a species requiring immediate monitoring of its populations. Targeted population studies have not been conducted in Mordovia yet. In Russia, this species is one of the widely distributed orchids (Efimov 2020). *Dactylorhiza fuchsii* is included in almost one third (28.8%) of the Red Data Books of Russian regions (Khapugin et al. 2020).

Figure 5. *Cypripedium calceolus* (photo A. Khapugin).

Figure 6. *Cypripedium guttatum*: a – habitus, b – flower (photo G. Chugunov).

Dactylorhiza incarnata (L.) Soó

Dactylorhiza incarnata (L.) Soó locations occur in 31 grid cells (10.4%) (Figure 2F). In Mordovia, this orchid species inhabits wet and marshy meadows and swamps (Silaeva et al. 2010). It is included in the Additional List of Red Data Book of the Republic of Mordovia (Silaeva 2017) as a species requiring immediate monitoring of its populations. *Dactylorhiza incarnata* subsp. *cruenta* (O.F. Müll.) P.D. Sell, on the other hand, is included in the Red Data Book of the Republic of Mordovia (rarity category 4, uncertain species) (Silaeva 2017) and was recently assessed as a Critically Endangered taxon (Khapugin et al. 2017a). Undoubtedly, this subspecies should be excluded from the next edition of the regional Red Data Book. Targeted population studies have not been conducted in Mordovia yet. In Russia, *D. incarnata* is included in 33.9% of the regional Red Data Books, while *D. incarnata* subsp. *cruenta* is included in 28.8% (Khapugin et al. 2020).

Dactylorhiza maculata (L.) Soó

Dactylorhiza maculata (L.) Soó occurs in 7 grid cells (2.3%) (Figure 2G). This species is legally protected in the Republic of Mordovia. *Dactylorhiza maculata* is included in the Red Data Book of the Republic of Mordovia (rarity category 2, vulnerable species) (Silaeva 2017). Recently, it has been classified as an Endangered taxon, according to the IUCN Red List categories (Khapugin et al. 2017a). Data on the overall abundance of this orchid in the region are lacking; it is probably low. Only one population-based study has been conducted for a small site of this species (Khapugin et al. 2015), which does not allow conclusions to be drawn about the status of all populations in Mordovia. In Russia, this species is one of the most widespread orchids (Efimov 2020). *Dactylorhiza maculata* is included in almost one third (30.5%) of regional Red Data Books in Russia (Khapugin et al. 2020).

Dactylorhiza viridis (L.) R. M. Bateman, Pridgeon & M. W. Chase

Dactylorhiza viridis (L.) R. M. Bateman, Pridgeon & M. W. Chase occurs in one grid cell (0.3%) (Figure 2H). This species was included in the Red Data Book of the Republic of Mordovia (Silaeva 2017) and in the synopsis of the regional flora (Silaeva et al. 2010). Here, the correct name is given in accordance with the international database (POWO 2020), the European source (Tatarenko et al. 2020) and the current survey of Russian orchids (Efimov 2020). *Dactylorhiza viridis* is legally protected in the region, it is included in the Red Data Book of the Republic of Mordovia (rarity category 2, vulnerable species) (Silaeva 2017). Recently, this species has been assessed as a Critically Endangered taxon, according to the IUCN Red List categories (Khapugin et al. 2017a). There are no targeted population-based studies conducted in Mordovia yet. In Russia, this species is one of the most widespread orchids, occurring in 71 regions (Efimov 2020). *Dactylorhiza viridis* is present in almost half (40.7%) of the Red Data Books of Russian regions (Khapugin et al. 2020). There is an urgent need to find the only population known in Mordovia currently and to monitor it annually.

Epipactis helleborine (L.) Crantz

Epipactis helleborine (L.) Crantz occurs in 53 grid cells (17.2%) (Figure 2I). In Mordovia, this orchid species colonizes various forest types and meadows (Silaeva et al. 2010). Most likely, this orchid is the most common orchid species in Mordovia. However, additional data are needed to clarify the actual distribution of this species. Targeted population-based studies have not been conducted here yet. In Russia, this species is one of the most widespread orchids, occurring in 68 regions (Efimov 2020). *Epipactis helleborine* is included in almost one third (30.5%) of Russia's regional Red Data Books (Khapugin et al. 2020).

Epipactis palustris (L.) Crantz

Epipactis palustris (L.) Crantz (Figure 7a) occurs in 6 grid cells (1.9%) (Figure 2J). This species is legally protected in the Republic of Mordovia. It is included in the Red Data Book of the Republic of Mordovia (rarity category 3, rare species) (Silaeva 2017). Recently, *E. palustris* has been assessed as an Endangered taxon, according to the IUCN Red List categories (Khapugin et al. 2017a). There are both high and low abundance populations in this region. *Epipactis palustris* is relatively resistant to anthropogenic pressures. Only one targeted regional study was conducted, which considered the largest *E. palustris* population in Mordovia (Khapugin et al. 2016b), where the population density of *E. palustris* was 12–50 individuals per 1 m^2. In Russia, this species is considered one of the most widespread orchids known for 65 regions (Efimov 2020). *Epipactis palustris* is included in most (69.5%) of the Red Data Books of Russian regions (Khapugin et al. 2020). The most important issue for future studies is to determine the total number of *E. palustris* populations in Mordovia.

Epipogium aphyllum Sw.

Epipogium aphyllum Sw. occurs in 3 grid cells (1.0%) (Figure 2K). This species is legally protected. It is included in the Red Data Book of the Republic of Mordovia (rarity category 1, endangered species)

(Silaeva 2017). Recently, this species has been classified as a Critically Endangered taxon, according to the IUCN Red List categories (Khapugin et al. 2017a). The overall abundance of this orchid in the region is always low (a few specimens per site).

Figure 7. Orchids of the Republic of Mordovia: a – *Epipactis palustris*, b – *Herminium monorchis*, c – *Orchis militaris*, d – *Ponerorchis cucullata* (photos A. Khapugin).

There is a lack of data on the population status of *E. aphyllum* due to the peculiarities of its biology. In Russia, this species is one of the more widely distributed orchids, occurring in 52 regions (Efimov 2020). *Epipogium aphyllum* is included in 66.1% of the Red Data Books of Russian regions (Khapugin et al. 2020). There is an urgent need to find and monitor currently the only population known in Mordovia.

Goodyera repens (L.) R.Br.

Goodyera repens (L.) R.Br. locations occur in 9 grid cells (2.9%), only in the western part of Mordovia (Figure 2L). This species is legally protected in Mordovia. It is included in the Red Data Book of Mordovia (rarity category 2, vulnerable species) (Silaeva 2017). Recently, this species has been assessed as an Endangered taxon, according to the IUCN Red List categories (Khapugin et al. 2017a). In the Mordovia State Nature Reserve, populations are stable and characterized by a high abundance. On other sites, the abundance is significantly lower. Despite the relatively high number of local populations in the Mordovia State Nature Reserve and outside, population studies have not been conducted in Mordovia yet. In Russia, this species is one of the most widespread orchids. It is registered in 69 Russian regions (Efimov 2020). *Goodyera repens* is included in almost one third (28.8%) of the regional Red Data Books of Russia (Khapugin et al. 2020). The most important issue for future studies is to determine the total number of *G. repens* populations in Mordovia.

Gymnadenia conopsea (L.) R.Br.

Gymnadenia conopsea (L.) R.Br. occurs in 9 grid cells (2.9%) (Figure 3A). This species is legally protected, being included in the Red Data Book of the Republic of Mordovia (rarity category 3, rare species) (Silaeva 2017). Recently, this species has been assessed as a Vulnerable taxon, according to the IUCN Red List categories (Khapugin et al. 2017a). Most populations are represented by a few individuals, although there are some populations inhabiting 200–300 m^2. Despite the lack of data on the population status of this orchid, no studies have been conducted in Mordovia yet. In Russia, this species is one of the most widespread orchid, occurring in 74 regions (Efimov 2020). *Gymnadenia conopsea* is included in half (49.2%) of the regional Red Data Books of Russia (Khapugin et al. 2020). We recommend the need for population-based studies of *G. conopsea* in Mordovia.

Hammarbya paludosa (L.) Kuntze

Hammarbya paludosa (L.) Kuntze occurs in one grid cell (0.3%) (Figure 3B). This species is legally protected in the Republic of Mordovia, it is included in the Red Data Book of the Republic of Mordovia (rarity category 1, endangered species) (Silaeva 2017). Previously, this species was assessed as Critically Endangered taxon according to the IUCN Red List categories (Khapugin et al. 2017a). No targeted population-based studies have been conducted, although only one regular population is known in Mordovia. In Russia, this species is one of the most widespread orchids, occurring in 56 regions (Efimov 2020). *Hammarbya paludosa* is included in more than half (62.7%) of the regional Red Data Books of Russia (Khapugin et al. 2020).

Herminium monorchis (L.) R.Br.

Herminium monorchis (L.) R.Br. (Figure 7b) occurs in 6 grid cells (1.9%) (Figure 3C). This species is legally protected in the region. It is included in the Red Data Book of the Republic of Mordovia (rarity category 2, vulnerable species) (Silaeva 2017). Recently, *H. monorchis* has been classified as a Critically Endangered taxon, according to the IUCN Red List categories (Khapugin et al. 2017a). There is a lack of data on the overall abundance of this orchid in the region, which is, however, probably quite low, due to the high vulnerability of this species to grazing (Silaeva 2017). Only one population study has been conducted for the most viable population of this species (Khapugin et al. 2015). However, it does not allow assessing the status of all *H. monorchis* populations in Mordovia. In Russia, this species is one of the most widespread orchids, being recorded in 62 regions (Efimov 2020). *Herminium monorchis* is included in almost half (40.7%) of the Red Data Books of Russian regions (Khapugin et al. 2020). The most important issue for future studies is to determine the total number of *H. monorchis* populations in Mordovia.

Malaxis monophyllos (L.) Sw.

Malaxis monophyllos (L.) Sw. occurs in 5 grid cells (1.6%) (Figure 3D). This species is legally protected in the Republic of Mordovia. It is included in the Red Data Book of the Republic of Mordovia (rarity category 1, endangered species) (Silaeva 2017). Previously, this species was assessed as an Endangered taxon, according to the IUCN Red List categories (Khapugin et al. 2017a). Data on the total occurrence of this orchid in the Republic of Mordovia are lacking. At all sites known so far, this orchid is represented by a few specimens, although its abundance is relatively stable over a long observation period (Silaeva 2017). The only population study was conducted for one of the most viable populations (Khapugin et al. 2016b). This does not allow conclusions to be drawn about the status of all populations in Mordovia, including the Mordovia State Nature Reserve, where most of the populations are found. In Russia, this species is one of the most widespread orchids, occurring in 57 regions (Efimov 2020). *Malaxis monophyllos* is included in almost half (44.1%) of Russia's regional Red Data Books (Khapugin et al. 2020). We advocate the need for population-based studies of *M. monophyllos* in Mordovia.

Neotinea ustulata (L.) R.M. Bateman, Pridgeon & M.W. Chase

Neotinea ustulata (L.) R.M. Bateman, Pridgeon & M.W. Chase occurs in one grid cell or 0.3% (Figure 3E). This species is given as *Orchis ustulata* L. in the Red Data Book of the Republic of Mordovia (rarity category 0, probably extinct species) (Silaeva 2017) and in the synopsis of the regional flora (Silaeva et al. 2010). Here it is correctly cited according to both international (POWO 2020) and national (Efimov 2020) sources. *Neotinea ustulata* is legally protected in the Republic of Mordovia. Recently, this species has been assessed as a Critically Endangered taxon, according to the IUCN Red List categories (Khapugin et al. 2017a). No data are available on either the abundance and population status nor the occurrence of this orchid species in Mordovia. For the last time this species was found in Mordovia more than 130 years ago (Silaeva 2017). Therefore, we can assume that *N. ustulata* has to be classified as a regionally extinct species, although this assumption should be confirmed/denied after targeted studies of the single site. *Neotinea ustulata* is included in more than one third (37.3%) of the Red Data Books of Russian regions (Khapugin et al. 2020). Currently, the rediscovery of *N. ustulata* in Mordovia or the detection of its absence in the regional flora is the most important issue.

Neottia cordata (L.) Rich.

Neottia cordata (L.) Rich. occurs in 2 grid cells (0.6%) (Figure 3F). This species is cited as *Listera cordata* (L.) R.Br. in the Red Data Book of the Republic of Mordovia (Silaeva 2017) and the synopsis of the regional flora (Silaeva et al. 2010). Here the correct name is given according to Efimov (2020) and POWO (2020). This species is legally protected being included in the Red Data Book of the Republic of Mordovia (rarity category 1, endangered species) (Silaeva 2017). Recently, this species has been classified as a Critically Endangered taxon, according to the IUCN Red List categories (Khapugin et al. 2017a). There is a lack of data on the abundance and population status of *N. cordata*, as its sites have not been rediscovered since the early 1990s, i.e., no targeted population-based studies have been conducted in Mordovia. However, the location of the two known populations within the Mordovia State Nature Reserve guarantees their protection against human impact. In Russia, this species is one of the most widespread orchids, recorded in 56 regions (Efimov 2020). *Neottia cordata* is included in almost half (44.1%) of the Red Data Books of Russian regions (Khapugin et al. 2020). Currently, the rediscovery of *N. cordata* in Mordovia is necessary with consistently initiated population monitoring.

Neottia nidus-avis (L.) Rich.

Neottia nidus-avis (L.) Rich. occurs in 30 grid cells (9.7%) (Figure 3G). This orchid species inhabits broadleaved and mixed forests in Mordovia (Silaeva et al. 2010). Although it is a relatively common species (Silaeva et al. 2010), some large gaps in its distribution in Mordovia are evident (Figure 3G). Probably, this species is currently one of the most common orchids in Mordovia. However, new floristic records are needed to fill the gaps in its distribution. Targeted population-based studies have not been conducted in the Republic of Mordovia yet. In Russia, this species is one of the most widespread orchids, occurring in 63 regions (Efimov 2020). *Neottia nidus-avis* is included in more than one third (37.3%) of the Red Data Books of Russian regions (Khapugin et al. 2020).

Neottia ovata (L.) Bluff & Fingerh.

Neottia ovata (L.) Bluff & Fingerh. occurs in 10 grid cells (3.2%) (Figure 3H). This species was mentioned in the synopsis of the regional flora (Silaeva et al. 2010) as *Listera ovata* (L.) R.Br. Here we give its correct name in accordance with international and national sources (Efimov 2020; POWO 2020). Unfortunately, this species is not legally protected in Mordovia, although there are few known locations of this orchid. This is particularly important as *N. ovata* is probably threatened by the disturbance of the hydrological regime of habitats. It prefers exclusively wet habitats, such as swampy deciduous and mixed forests, afforested ravines, edges of mires and swampy forests (Silaeva et al. 2010). Targeted population-based studies have not been conducted in Mordovia yet. In Russia, this species is one of the widely distributed orchids. It is registered in 67 regions (Efimov 2020). *Neottia ovata* is included in half (49.2%) of the regional Red Data Books of Russia (Khapugin et al. 2020). We advocate the need for population-based studies of *N. ovata* in Mordovia, with special attention to granting conservation status.

***Orchis militaris* L.**

Orchis militaris L. (Figure 7c) occurs in 7 grid cells (2.3%) (Figure 3I). This species is legally protected in the Republic of Mordovia. It is included in the Red Data Book of the Republic of Mordovia (rarity category 1, endangered species) (Silaeva 2017). Recently, this species has been assessed as a Critically Endangered taxon, according to the IUCN Red List categories (Khapugin et al. 2017a). A single population study was conducted for one of the most viable populations in the region (Khapugin et al. 2016b). However, there is a need to study other populations of this orchid in Mordovia as well, to obtain a complete overview of the status of *O. militaris* in Mordovia. In Russia, this species is one of the most widespread orchids, occurring in 62 regions (Efimov 2020). *Orchis militaris* is included in most (76.3%) of the Red Data Books of Russian regions (Khapugin et al. 2020). The most important question for future studies is to determine the total number of *O. militaris* populations in Mordovia.

***Platanthera bifolia* (L.) Rich.**

Platanthera bifolia (L.) Rich. sites occur in 56 grid cells (17.2%) (Figure 3J). In Mordovia, this orchid species inhabits light forests and forest glades (Silaeva et al. 2010). Apparently, *P. bifolia* is the most common orchid in Mordovia, together with *Epipactis helleborine*, *Dactylorhiza incarnata* and, probably, *Neottia nidus-avis*. However, more data are needed to clarify the current distribution of *P. bifolia* in Mordovia. Targeted population studies were conducted for two years in the Mordovia State Nature Reserve, and the morphometric parameters of this orchid were analyzed. This species has a certain degree of resistance to moderate anthropogenic pressure and needs a certain level of illumination for flowering initiation (Khapugin and Korochkina 2017, 2019). In Russia, this species is one of the most widespread orchids, occurring in 69 regions (Efimov 2020). *Platanthera bifolia* is included in more than one third (37.3%) of the Red Data Books of Russian regions (Khapugin et al. 2020). The most important issue for future studies is to determine the total number of *P. bifolia* populations in Mordovia.

Platanthera chlorantha (Custer) Rchb.

Platanthera chlorantha (Custer) Rchb. occurs in 13 grid cells (4.2%) (Figure 3K). In Mordovia, this orchid species inhabits deciduous and mixed forests (Silaeva et al. 2010). It is included in the additional list of Red Data Book of the Republic of Mordovia (Silaeva 2017), as a species calling for population monitoring. Targeted population-based studies have not been conducted in Mordovia yet. *Platanthera chlorantha* is included in one third (32.2%) of the Red Data Books of Russian regions (Khapugin et al. 2020). The most important issue for future studies is the assessment of the total number of *P. chlorantha* populations in Mordovia with special attention to the influence of certain environmental factors. The latter is especially important considering the recent data on the significant dependence of *P. chlorantha* population status on habitat conditions in the Greater Caucasus (Mursal et al. 2020).

Ponerorchis cucullata (L.) X.H. Jin, Schuit. & W.T. Jin

Ponerorchis cucullata (L.) X.H. Jin, Schuit. & W.T. Jin (Figure 7d) occurs in 10 grid cells (3.2%) (Figure 3L). This species was mentioned as *Neottianthe cucullata* (L.) Schltr. in the Red Data Book of the Republic of Mordovia (Silaeva 2017) and the synopsis of the regional flora (Silaeva et al. 2010). We give here the correct globally accepted name (Efimov 2020; POWO 2020). This species is legally protected in the Republic of Mordovia. It is included in the Red Data Book of the Republic of Mordovia (rarity category 2, vulnerable species) (Silaeva 2017). Recently, this species has been classified as an Endangered taxon, according to the IUCN Red List categories (Khapugin et al. 2017a). This species is one of the most important orchid species in Mordovia. Previously, four of all known *P. cucullata* populations in Mordovia have been studied for more than ten years: two populations in Mordovia State Nature Reserve, one in Smolny National Park and one in southeastern Mordovia (Khapugin et al. 2016a). Earlier studies were also conducted by the staff of the Mordovia State Nature Reserve (Sanaeva 2011). According to these studies, most viable populations of this orchid are located in the Mordovia State Nature Reserve. In Russia, this species is one of the most widespread orchids, registered in 47 regions (Efimov 2020). *Ponerorchis cucullata* is included in more than half (66.1%) of the regional Red Data Books of Russia (Khapugin et al. 2020). The most important issue for future studies is the continuation of population studies of *P. cucullata* in Mordovia.

Conservation of Orchids in the Republic of Mordovia

Most orchids are rare, and threatened species, both regionally and globally (Cribb et al. 2003; Juiling et al. 2020; Kottawa-Arachchi and Gunasekara 2020; Popovich et al. 2020), with some exceptions (Besi et al. 2020). Threats to orchids are predominantly associated with features of their biology (Seaton et al. 2013; Shefferson et al. 2020). Comprehensive studies of the ecology and biology of these threatened plants are needed to successfully protect natural orchid populations. In Mordovia, most orchids are defined as threatened taxa (Figure 8), while seven taxa still have "Not Evaluated" status. Detailed studies of their populations and distribution are ultimately needed according to the taxa assessment criteria proposed by the IUCN Red List.

Despite a high proportion of threatened taxa, population-based monitoring has been conducted for only a few species, including long-term observations of the Endangered *Cypripedium calceolus* (Khapugin et al. 2017c) and the Endangered *Ponerorchis cucullata* (Khapugin et al. 2016a), and short-term studies of the Critically Endangered *Corallorhiza trifida* (Khapugin et al. 2015), The Endangered *Dactylorhiza maculata* (Khapugin et al. 2015), Endangered *Epipactis palustris* (Khapugin et al. 2016b), Critically Endangered *Herminium monorchis* (Khapugin et al. 2015), Endangered *Malaxis monophyllos* (Khapugin et al. 2016b), Critically Endangered *Orchis militaris* (Khapugin et al. 2016b), and Not Evaluated (probably, Least Concern) *Platanthera bifolia* (Khapugin and Korochkina 2017, 2019). Currently, there are no reliable data on the population status of the other 15 orchid species in Mordovia.

There is an urgent need to establish a long-term orchid population monitoring program aimed at understanding the population status and abundance, and thus population trends, to prevent orchid extinction. The use of modern taxonomic treatments instead of "traditional" (= outdated) or national/sub-national sources that introduce confusion and misunderstanding into data presentation, similar to specialized vocabulary (jargon), is another important issue that should be considered in future studies (e.g., Martínez and Mammola 2020).

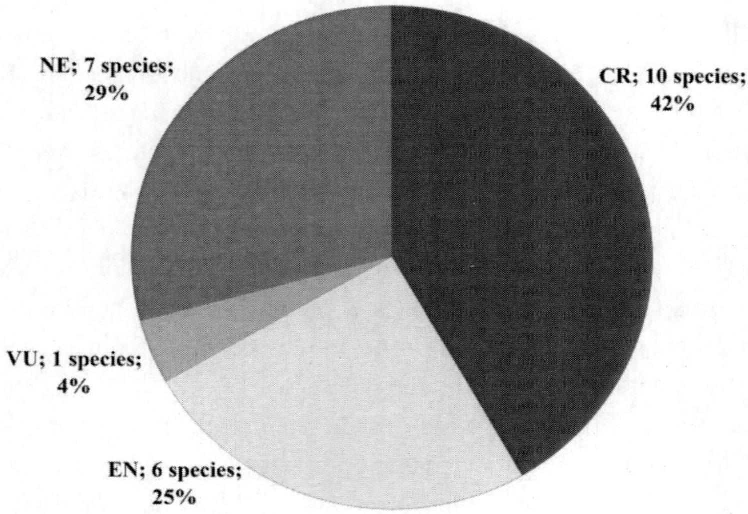

Figure 8. The regional IUCN Red List status of orchid species known in the Republic of Mordovia (Russia). Designations: CR – Critically Endangered, EN – Endangered, VU – Vulnerable, NE – Not Evaluated.

CONCLUSION

According to the summarized and standardized data, 24 orchid species are known in Mordovia. Two regional hotspots of orchid diversity were found, located in the Mordovia State Nature Reserve and near the Mordovia State University Biological Station. These hotspots allow multiple orchid species to be studied in a limited area. We call for the following actions: i) estimation of the seven orchid species currently classified as "Not Evaluated" according to IUCN Red List guidelines; ii) development and maintenance of a long-term program for monitoring orchid populations in Mordovia; iii) use of modern international taxonomic databases, such as POWO (2020), by regional researchers to keep pace with global changes in plant nomenclature and taxonomy.

References

Acharya, Kamal P., Ole R. Vetaas, and H. J. B. Birks. "Orchid Species Richness along Himalayan Elevational Gradients: Orchid Richness along Elevational Gradients." *Journal of Biogeography* 38, no. 9 (2011): 1821–33.

Ageeva, Anna M., Evgeniy A. Belyakov, Olga V. Biryukova, Viktoria V. Bondareva, Vladimir M. Vasjukov, Vasiliy S. Vishnyakov, Margarita M. Gafurova, Tatyana V. Gorbushina, Oleg G. Grishutkin, Evgeniy G. Zibzeev, Alexander V. Ivoilov, Valentina N. Ilyina, Nadezhda V. Koneva, Andrey Yu. Korolyuk, Anna E. Mitroshenkova, Lyubov A. Novikova, Sergey V. Saksonov, Stepan A. Senator, Larisa V. Sidyakina, Tatyana B. Silaeva, Maria V. Sosnina, Ekaterina S. Fedashova, Anatoliy A. Khapugin, Lyudmila A. Khudoykina, Anna A. Shestakova, Alexander A. Shkulev, Gennadiy G. Chugunov, and Natalya A. Yuritsyna. *Conservation of rare species of plants and fungi in the Volga River basin: Series Floristic Yearbook, 2019* (Togliatti: Anna, 2020). [in Russian].

Barberena, Felipe Fajardo Villela Antolin, Felipe Fajardo Villela, José Fernando A. Baumgratz, and Fábio de Barros. "Ecological Data for an Orchid Diversity Hotspot Show That the Subtribe Laeliinae May Be Endangered in the Brazilian Atlantic Forest." *Nordic Journal of Botany* 36, no. 7 (2018): e01728.

Bernardos, Sonia, Daniel Tyteca, Mónica García-Barriuso, Antonio Crespí, Adriano Castro, and Francisco Amich. "Current Status and Conservation of the Lusitan Duriensean Orchids (Duero Basin, NE Portugal and CW Spain)." *Acta Botanica Gallica* 153, no. 3 (2006): 273–84.

Besi, Edward E., Universiti Putra Malaysia, Dome Nikong, Runi S. Pungga, Rusea Go, Kg Sungai Tong, Terengganu, Malaysia, Forest Department Sarawak, and Universiti Putra Malaysia. "Wild Orchid Diversity of Highland Forest in the Heart of Borneo: Long Banga and Tama Abu, Sarawak." *Nature Conservation Research* 5, no. Suppl.1 (2020). https://doi.org/10.24189/ncr.2020.048.

Calevo, Jacopo, and Miriam Bazzicalupo. "Less Is More: Low-Cost in Vitro Propagation of an Endangered Italian Orchid." *Nature Conservation Research* 5, no. Suppl.1 (2020). https://doi.org/10.24189/ncr.2020.043.

Christenhusz, Maarten J. M., and James W. Byng. "The Number of Known Plants Species in the World and Its Annual Increase." *Phytotaxa* 261, no. 3 (2016): 201–17.

Crain, Benjamin J., and Raymond L. Tremblay. "Do Richness and Rarity Hotspots Really Matter for Orchid Conservation in Light of Anticipated Habitat Loss?" *Diversity & Distributions* 20, no. 6 (2014): 652–62.

Cribb, Phillip J., Shelagh P. Kell, Kingsley W. Dixon, and Russell L. Barrett. "*Orchid conservation: a global perspective*." In *Orchid conservation*, edited by Kingsley W. Dixon, Shelagh P. Kell, Russell L. Barrett, and Phillip J. Cribb, 1–24, Kota Kinabalu: Natural History Publications, 2003.

de la Torre Llorente, Daniel. "Conservation Status of the Family Orchidaceae in Spain Based on European, National, and Regional Catalogues of Protected Species." *The Scientific World Journal* (2018). doi: 10.1155/2018/7958689.

Djordjević, Vladan, and Spyros Tsiftsis. "The Role of Ecological Factors in Distribution and Abundance of Terrestrial Orchids." In *Orchids Phytochemistry, Biology and Horticulture, Reference Series in*

Phytochemistry. Fundamentals and Applications, edited by Jean-Michel Merillon, and Hippolyte Kodja, 1–71. Cham: Springer International Publishing, 2020, doi:10.1007/978-3-030-11257-8_4-1.

Djordjević, Vladan, Spyros Tsiftsis, Dmitar Lakušić, Slobodan Jovanović, and Vladimir Stevanović. "Factors Affecting the Distribution and Abundance of Orchids in Grasslands and Herbaceous Wetlands." *Systematics and Biodiversity* 14, no. 4 (2016a): 355–70.

Djordjević, V., S. Tsiftsis, D. Lakušić, and V. Stevanović. "Niche Analysis of Orchids of Serpentine and Non-Serpentine Areas: Implications for Conservation." *Plant Biosystems* 150, no. 4 (2016b): 710–19.

Djordjević, Vladan, Dmitar Lakušić, Slobodan Jovanović, and Vladimir Stevanović. "Distribution and Conservation Status of Some Rare and Threatened Orchid Taxa in the Central Balkans and the Southern Part of the Pannonian Plain." *Wulfenia* 24 (2017): 143–62.

Efimov, Petr G. "Orchids of Russia: Annotated Checklist and Geographic Distribution." *Nature Conservation Research* 5, no. Suppl.1 (2020): 1–18. https://doi.org/10.24189/ncr.2020.018.

Egorova, Natalya Yu; Venera N. Suleimanova, and Tatyana L. Egoshina. "The Dynamics of the Demographic Structure of Populations of *Cypripedium calceolus* L. (Orchidaceae) in the Valley of the River Vyatka." *Bulletin of Moscow Society of Naturalists* 125, no. 2 (2020): 51–59. [in Russian].

Eliáš, Pavol, Daniel Dítě, Ján Kliment, Richard Hrivnák, and Viera Feráková. "Red List of Ferns and Flowering Plants of Slovakia, 5th edition (October 2014)." *Biologia* 70, no. 2, (2015): 218–28.

Fateryga, Alexander V., Anton V. Popovich, Valentina V. Fateryga, Karel C. A. J. Kreutz. "*Cephalanthera epipactoides* (Orchidaceae) in Russia." *Nature Conservation Research* 5, no. Suppl.1 (2020): 69–76. https://doi.org/10.24189/ncr.2020.014.

Gale, Stephan W., Gunter A. Fischer, Phillip J. Cribb, and Michael F. Fay. "Orchid Conservation: Bridging the Gap Between Science and

Practice." *Botanical Journal of the Linnean Society* 186, no. 4 (2018): 425–34.

Givnish, Thomas J., Daniel Spalink, Mercedes Ames, Stephanie P. Lyon, Steven J. Hunter, Alejandro Zuluaga, Alfonso Doucette, et al. "Orchid Historical Biogeography, Diversification, Antarctica and the Paradox of Orchid Dispersal." *Journal of Biogeography* 43, no. 10 (2016): 1905–16.

Gravendeel, Barbara, Ann Smithson, Ferry J. W. Slik, and Andre Schuiteman. "Epiphytism and Pollinator Specialization: Drivers for Orchid Diversity?" *Philosophical Transactions of the Royal Society of London. Series B, Biological Sciences* 359, no. 1450 (2004): 1523–35.

Grishutkin, Oleg G. "Possibility of using of GIS in botanical researches." *Mordovia University Bulletin* 3–4 (2013): 16–20. [in Russian].

Halme, Panu, Saija Kuusela, and Aino Juslén. "Why Taxonomists and Ecologists Are Not, but Should Be, Carpooling?" *Biodiversity and Conservation* 24, no. 7 (2015): 1831–36.

Hammer, Øyvind, David A. T. Harper, and Paul D. Ryan. "PAST: Paleontological Statistics Soft-ware Package for Education and Data Analysis." *Palaeontologia Electronica* 4, no. 1 (2001): 9.

Huang, Bao-Qiang, Xiao-Qin Yang, Fei-Hai Yu, Yi-Bo Luo, and Yun-Dong Tai. "Surprisingly High Orchid Diversity in Travertine and Forest Areas in the Huanglong Valley, China, and Implications for Conservation." *Biodiversity and Conservation* 17, no. 11 (2008): 2773–86.

Ječmenica, Vladimir, Vincent Droissart, Nausicaa Noret, and Tariq Stévart. "Taxonomy of Atlantic Central African Orchids 5. A New Species of *Angraecum* Sect. *Conchoglossum* (Orchidaceae, Angraecinae) from Gabon and Cameroon." *PhytoKeys* 61, no. 61 (2016): 61–71.

Juiling, Suzika, Scholena K. Leon, Jamirus Jumian, Sandy Tsen, Yew Leung Lee, Eyen Khoo, John B. Sugau, Reuben Nilus, Joan T. Pereira, Alviana Damit, Sasikumar Tanggaraju, Peter O'Byrne, Sukaibin Sumail, Handry Mujih, and Colin R. Maycock,

"Conservation Assessment and Spatial Distribution of Endemic Orchids in Sabah, Borneo." *Nature Conservation Research* 5, no. Suppl.1 (2020). https://doi.org/10.24189/ncr.2020.053.

Khapugin, Anatoliy A. "A Global Systematic Review on Orchid Data in Protected Areas." *Nature Conservation Research* 5, no. Suppl.1 (2020): 19–33. https://doi.org/10.24189/ncr.2020.019.

Khapugin, Anatoliy A., and Tatyana B. Silaeva. "The Arrangement of Threatened Plants in Mordovia: The Role of Biodiversity Research Centers." *Ecoscience* 27, no. 3 (2020): 157–64.

Khapugin, Anatoliy A., and Anna M. Korochkina. "Population-based Study of *Platanthera bifolia* in the Mordovia State Nature Reserve (Republic of Mordovia, Central Russia)." *Orchidee Spontanee d'Europa – European Native Orchids* 60, no. 2 (2017): 276–94.

Khapugin, Anatoliy A., and Anna M. Korochkina. "About *Platanthera bifolia* (Orchidaceae) Population in the Mordovia State Nature Reserve (Russia)." *Proceedings of the Mordovia State Nature Reserve* 23 (2019): 195–209. [in Russian].

Khapugin, Anatoliy A., Galina V. Andryushechkina, and Gennadiy G. Chugunov. "About Status of Populations of Plant Species from Red Data Book of Russian Federation in the Mordovia State Nature Reserve in 2011." *Proceedings of the Mordovia State Nature Reserve* 10 (2012): 321–27. [in Russian].

Khapugin, A. A., A. A. Semchuk, T. B. Silaeva, and G. G. Chugunov. "A Comparative Characteristics of *Cypripedium calceolus* L. (Orchidaceae, Monocotyledones) Coenopopulations in the Republic of Mordovia." *Povolzhskiy Journal of Ecology* 3 (2014): 403–10. [in Russian].

Khapugin, Anatoliy A., Anastasia A. Semchuk, Maria V. Sosnina, Gennadiy G. Chugunov, Tatyana B. Silaeva, and Elena V. Vargot. "Biomorphology of Five Rare Orchid Species (Orchidaceae Juss.) in Populations of Central Russia.". *Proceedings of the Mordovia State Nature Reserve* 15 (2015): 194–205. [in Russian].

Khapugin, Anatoliy A., Gennadiy Chugunov, Tatyana B. Silaeva, and Elena N. Kunaeva. "*Neottianthe cucullata* (L.) Schltr. (Orchidaceae

Juss.), an Endangered Orchid in Central Russia." *Wulfenia* 23 (2016a): 189–202.

Khapugin, Anatoliy A., Tatyana B. Silaeva, Anastasia A. Semchuk, and Elena N. Kunaeva. "Populations of *Orchis militaris* L., *Epipactis palustris* (L.) Crantz and *Malaxis monophyllos* (L.) Sw. in the Republic of Mordovia (Central Russia)." *Biodiversity: Research and Conservation* 42 (2016b): 33–40.

Khapugin, Anatoliy A., Tatyana B. Silaeva, Elena V. Vargot, Gennadiy G. Chugunov, Galina A. Grishutkina, Oleg G. Grishutkin, Elena V. Pismarkina, and Julia S. Orlova. "Estimation of Taxa Included in the First Volume of the Red Data Book of the Republic of Mordovia (Russia) Using the IUCN Red List Categories and Criteria." *Nature Conservation Research* 2, no. Suppl.1 (2017a): 164–89. doi: 10.24189/ncr.2017.004 [in Russian].

Khapugin, Anatoliy A., Tatyana B. Silaeva, Elena V. Vargot, and Gennadiy G. Chugunov. "IUCN Guidelines Using for Assessment of Plants from the Red Book of Russian Federation at Regional Level: A Case Study for the Republic of Mordovia (Russia)." *Hacquetia* 16, no. 1 (2017b): 19–33.

Khapugin, Anatoliy A., Gennadiy G. Chugunov, and Elena V. Vargot. "*Cypripedium calceolus* (Orchidaceae) in Central Russia: A Case Study for Its Populations in Two Protected Areas in the Republic of Mordovia (Russia)." *Lankesteriana* 17, no. 3 (2017c): 403–17. https://doi.org/10.15517/lank.v17i3.31577.

Khapugin, Anatoliy A., Tatyana B. Silaeva, and Lyudmila A. Khudoykina. "A Record of *Cypripedium calceolus* in Urban Ecosystem of Culture Park of the Komsomolskiy Settlement, Republic of Mordovia, Central Russia." *MIOS Journal* 18, no. 9 (2017d): 2–9.

Khapugin, Anatoliy A., Igor V. Kuzmin, and Tatyana B. Silaeva. "Anthropogenic Drivers Leading to Regional Extinction of Threatened Plants: Insights from Regional Red Data Books of Russia." *Biodiversity and Conservation* 29, no. 8 (2020): 2765–77.

Kirillova, Irina A., and Dmitry V. Kirillov. "Impact of Weather Conditions on Seasonal Development, Population Structure and Reproductive Success on *Dactylorhiza traunsteineri* (Orchidaceae) in the Komi Republic (Russia)." *Nature Conservation Research* 5, no. Suppl.1 (2020): 77–89. doi: 10.24189/ncr.2020.016.

Kottawa-Arachchi, J. Dananjaya, and Samantha Gunasekara, R. "Research Priorities and Future Directions in Conservation of Wild Orchids in Sri Lanka: A Review." *Nature Conservation Research* 5, no. Suppl.1 (2020): 34–45. doi: 10.24189/ncr.2020.029.

Kull, Tiiu, Ulvi Selgis, Miguel Villoslada Peciña, Mirjam Metsare, Aigi Ilves, Kadri Tali, Kalev Sepp, Kalevi Kull, and Richard P. Shefferson. "Factors Influencing IUCN Threat Levels to Orchids across Europe on the Basis of National Red Lists." *Ecology and Evolution* 6, no. 17 (2016): 6245–65.

Le Roux, Johannes J., Cang Hui, Maria L. Castillo, José M. Iriondo, Jan-Hendrik Keet, Anatoliy A. Khapugin, Frédéric Médail, Marcel Rejmánek, Genevieve Theron, Florencia A. Yannelli, and Heidi Hirsch. "Recent Anthropogenic Plant Extinctions Differ in Biodiversity Hotspots and Coldspots." *Current Biology* 29, no. 17 (2019): 2912–18.

Martínez, Alejandro, and Stefano Mammola. "Specialized Terminology Limits the Reach of New Scientific Knowledge." *bioRxiv* (2020). doi: 10.1101/2020.08.20.258996.

Mursal, Nigar, Naiba P. Mehdiyeva, and Aida G. Ibrahimova. "Population Status and Ecology of *Platanthera chlorantha* (Orchidaceae) in the Greater Caucasus (Azerbaijan)." *Nature Conservation Research* 5, no. Suppl.1 (2020): 114–24. doi: 10.24189/ncr.2020.046.

Perazza, Giorgio, and Michela Decarli. "Monitoring of *Cypripedium calceolus* (Orchidaceae) in the Adamello-Brenta Natural Park (Italy)." *Nature Conservation Research* 5, no. Suppl.1 (2020): 178–84. doi: 10.24189/ncr.2020.045.

Popovich, Anton V., Elena A. Averyanova, and Lev M. Shagarov. "Orchids of the Black Sea Coast of Krasnodarsky Krai (Russia):

Current State, New records, Conservation." *Nature Conservation Research* 5, no. Suppl.1 (2020): 46–68. doi: 10.24189/ncr.2020.047.

POWO. "*Plants of the World Online*." (2020). Available from http://powo.science.kew.org/.

Romano, Vito A., Leonardo Rosati, and Simonetta Fascetti. "Trends in Population Size of *Ophrys argolica* subsp. *biscutella* in the Appennino Lucano-Val d'Agri-Lagonegrese National Park (Italy)." *Nature Conservation Research* 5, no. Suppl.1 (2020): 155–64. doi: 10.24189/ncr.2020.058.

Sanaeva, Lyudmila V. "The Dynamics of Plant Communities and Groupings, Which Include Rare and Endangered Species of Vascular Plants." *Mordovia University Bulletin* 4 (2011): 56–89. [in Russian].

Seaton, Philip, Jonathan P. Kendon, Hugh W. Pritchard, Dwi Murti Puspitaningtyas, and Tim R. Marks, "Orchid Conservation: The Next Ten Years." *Lankesteriana* 13, no. 1–2 (2013): 93–101. doi: 10.15517/lank.v0i0.11545.

Shefferson, Richard P., Hans Jacquemyn, Tiiu Kull, and Michael J. Hutchings. "The Demography of Terrestrial Orchids: Life History, Population Dynamics and Conservation." *Botanical Journal of the Linnean Society* 192, no. 2 (2020): 315–32.

Shirokov, Alexander I., Vera V. Syrova, Alexey V. Salokhin, Ivan N. Markelov, Elena V. Andronova, and Elena V. Ganyushkina. "Conservation Issues and Infraspecific Polymorphism of *Cypripedium guttatum* on Selected Locations in Russia." *Nature Conservation Research* 5, no. Suppl.1 (2020): 145–54. doi: 10.24189/ncr.2020.054.

Silaeva, Tatyana B. ed., *Red Data Book of the Republic of Mordovia: in 2 vol., vol. 1: Rare species of plants and fungi, 2nd ed.* (Saransk: Publisher of the Mordovia State University, 2017). [in Russian].

Silaeva, Tatyana B. Igor V. Kiryukhin, Gennadiy G. Chugunov, Vasiliy K. Levin, Sergey R. Mayorov, Elena V. Pismarkina, Anna M. Ageeva, and Elena V. Vargot. *Vascular plants of the Republic of Mordovia (synopsis of flora)*. (Saransk: Publisher of the Mordovia State University, 2010). [in Russian].

Tatarenko, Irina, Michael Dodd, Hilary Wallace, Graham Bellamy, and Andy Fleckney. "Protecting Small Populations of Rare Species. Case Study on *Dactylorhiza viridis* (Orchidaceae) in Fancott Woods and Meadows SSSI, Bedfordshire, UK." *Nature Conservation Research* 5, no. Suppl.1 (2020): 165–71. doi: 10.24189/ncr.2020.028.

Terentieva, Elena I., Tatyana I. Varlygina, Galina F. Darman, Galina V. Degtjareva, Sergey V. Efimov, and Tagir H. Samigullin. "Revision and Distribution of *Liparis* Species (Orchidaceae) in Amur Region (Russia)." *Nature Conservation Research* 5, no. Suppl.1. (2020): 102–13. doi: 10.24189/ncr.2020.044.

The Plant List. *Version 1.1* (2013). Available from www.theplantlist.org/.

Tsiftsis, Spyros, Ioannis Tsiripidis, Vassiliki Karagiannakidou, and Dimitrios Alifragis. "Niche Analysis and Conservation of the Orchids of East Macedonia (NE Greece)." *Acta Oecologica* 33, no. 1. (2008): 27–35.

Wagensommer, Robert Philipp, Pietro Medagli, Alessio Turco, and Enrico V. Perrino. "IUCN Red List Evaluation of the Orchidaceae Endemic to Apulia (Italy) and Considerations on the Application of the IUCN Protocol to Rare Species." *Nature Conservation Research* 5, no. Suppl.1 (2020): 90–101. doi: 10.24189/ncr.2020.033

Yamashkin, Anatoliy A. ed. *Geographical atlas of Republic of Mordovia*. (Saransk: Publisher of the Mordovia State University, 2012). [in Russian].

In: Orchidaceae
Editor: Vladan Djordjević

ISBN: 978-1-68507-189-9

Chapter 5

INTERRELATIONSHIP BETWEEN THE ORCHID COMMUNITY AND ITS PHOROPHYTES IN CABO SAN ANTONIO, GUANAHACABIBES PENINSULA, CUBA

***Ernesto Mujica*[1,*]*, Andreu Bonet*[2]*, Elaine González*[1]*, Alejandro Camejo*[3] *and Josep Raventós*[2]

[1]Jardín Botánico Orquideario Soroa, Candelaria, Artemisa, Cuba
[2]Departamento de Ecología, Universidad de Alicante, Alicante, Spain
[3]Parque Nacional Guanahacabibes, Sandino, Pinar del Río, Cuba

ABSTRACT

This study was carried out in Cabo San Antonio, Guanahacabibes National Park, Cuba in 2019 and the aim was to analyse the relationship between the community of orchid species and their associated phorophytes to determine if specificity exists between the two. To determine the location of the orchid species and study their

[*] Corresponding Author's E-mail: emujica@uart.edu.cu.

interrelationship with the phorophyte community, transects were plotted in different plant formations at the site, or at least those deemed most important, such as sub-coastal xeromorphic scrub, semi-deciduous forest and coastal sand vegetation. In the areas under study, a total of 1090 individuals of 38 species of trees were counted, 26 of which harbored at least one epiphyte in 265 of the phorophyte total. Of the eight orchid species documented, 1244 individuals were counted. To characterize the relationship between epiphyte and phorophyte communities, a multivariate analysis was carried out on the frequency matrix of 26 species, including a category that included unidentified trees that had died. Our results suggest that the relationship between the epiphyte community and phorophyte species is generally asymmetric, such that there are generalist epiphyte species with little specificity of habitat or low requirements, and phorophytes that are also generalists and able to host several species of orchids and other epiphytes. Some orchid species were also found only on specific phorophyte species even in areas where other phorophyte species occurred nearby. We hypothesize that orchids may display specificity for certain types of mycorrhizal fungi needed for seed germination that are found only on the bark of these phorophyte species.

Keywords: biodiversity, Orchidaceae, host specificity, Cuba

INTRODUCTION

Epiphytic orchids differ in many respects from terrestrial orchids (Zotz and Schmidt 2006). For example, epiphytes do not have a stage underground, and their low density suggests that there is little intra- and inter-species competition (Zotz and Hietz 2001). In addition, the dynamics of the substrates (bark, branches, trunks and species) exert a very strong influence on the dynamics of epiphyte populations (Zotz and Schmidt 2006).

Although epiphytes form an important part of plant diversity, relatively little is known on the specificity of the relationship between these unique plants and their hosts. This association becomes especially interesting when taking into account the fact that epiphytes usually live together in clusters or are unevenly distributed (Trapnell and Hamrick 2006).

On a local scale, epiphytes partition forests horizontally depending on the density of micro-environments and the structural characteristics of the various hosts providing different opportunities for anchoring. In other words, a complete mosaic of micro-habitats, or a physical mosaic is provided by the environment for their growth (Martínez-Meléndez 2008). These micro-environments can be caused by differences in the availability of moisture, sunlight, or the presence of bryophytes (Krömer and Gradstein 2003; Krömer et al. 2007).

The age of the area available for colonization can affect the abundance of epiphytes and the composition of their communities (Catling et al. 1986; Pérez 1982). Moreover, the architecture of phorophytes and structure of their cover, ramification, leaf size, etc., can exert a strong influence on epiphyte germination and anchorage (Benzing 1979). Catling and Lefkovitch (1989) suggested that the DBH of phorophytes is an indicator of the approximate age of the tree species, which can be important in two ways: 1) the larger surface area, the larger area available for colonization, and 2) the seeds of epiphytic species have a higher probability of making physical contact with the host.

Although host preference has been discussed among epiphyte ecologists, we suggest that many studies of this kind have not been documented correctly. A challenge for testing host preference is that host species differ in abundance. It is expected that host species of greater abundance will support more epiphyte individuals (Vergara et al. 2010). However, few authors (Bennett 1987, Bernal et al. 2005, Martínez-Meléndez et al. 2008) have incorporated abundance of host species and a measurement of abundance of epiphyte species as a response variable in their analysis of host–epiphyte association. In host preference, epiphytes must be able to colonize all tree species but, for a few host species, their abundance must be greater than that expected by chance. In host specificity, an epiphyte species must occur on a single host species (Vergara et al. 2010).

Anchorage of orchid seeds onto the substrate depends on many biotic and abiotic factors. The most critical biotic factor is the presence of mycorrhizae in the substrate when they arrive (Trapnell and Hamrick 2006). The dependence on mycorrhizal fungi for germination is an important component of orchid biology (Arditti 1992) and may limit their distribution (Clements 1987), as well recruitment of new individuals in the populations.

Since it was found out that the availability of sites has a strong influence on the distribution of populations, and that those harbouring orchids are generally small and formed by patches or collections of individuals (Ackerman 1986), it has become extremely important for phorophyte species to be included in the development of orchid conservation strategies (Migenis and Ackerman 1993).

Few studies have been carried out in Cuba to analyse the communities of Cuban orchids and the ecosystems they inhabit, mainly because they have not been appropriately prioritized, coupled with a lack of people trained to undertake the task. Moreover, the distance at which the communities are found, and the objective conditions of researchers in making the studies, poses other obstacles. As a result, the application of strategies for handling and conserving populations of endangered orchid populations remains largely unknown for one reason or another.

Approximately 28 different species of orchid live in the Guanahacabibes National Park (GNP) in westernmost Cuba (Ferro et al. 1995), occupying almost all the plant formations in the region. However, the diversity of tree species means that most of them occupy well-defined habitats.

This paper proposes to analyse the relationship between the community of orchid species and that of phorophyte species in general in the GNP, to determine whether there is a phorophyte-orchid species specificity.

METHODS

The GNP lies on the Cabo Corrientes and Cabo San Antonio peninsulas. This study was carried out in the latter part of 2019. A total of 18 orchid species have been reported in Cabo San Antonio, 17 of which are epiphytes (Mújica, unpublished).

To determine the location of the orchid species and study their interrelationship with the phorophyte community, transects were plotted of varying lengths in different plant formations on the site, or at least the most important, such as sub-coastal xeromorphous scrub (SXS), semi-deciduous forest (SDF) and coastal sand vegetation (CSV). The length and width of the transects were consistent with the characteristics of each plant formation. Having analysed the information from the transects, 11 plots were planted in most crowded sites, to study their distribution pattern on phorophytes. The decision was made to look for uniformly dimensioned plots. As in many places of SXS is never much more that 20 m wide, plot measurements were set at 20 m × 20 m, or 400 m^2 each.

The phorophytes on each plot were mapped, regardless of the orchid species present or level of occupation. The characterisation only mapped phorophyte individuals with a DBH of 4 cm or more as epiphytes are not usually found on those with smaller diameters (Ferro 2004).

Statistical Analysis

For the analysis of the relationship of the species in the study with the phorophyte community in general, a Detrended Correspondence analysis (DCA) was chosen, to enable joint analysis of phorophytes and orchids in the space generated by the coordinate's axes (Ter Braak 1995). The analysis was run on CANOCO Version 4.5 February 2002 – a program written by Cajo J. F. Ter Braak (C) 1988-2002 Biometris - Quantitative Methods in the Life and Earth Sciences, Plant Research International, Wageningen University and Research Center. Because some authors point out that this method concerns frequencies (the counts of individual phorophytes with orchids), and that the data can be square-root transformed (Leps and Smilauer 2003), we opted not to use the transformation in this case, to be able to use the mathematical properties of the statistics as a variable measure.

RESULTS

The Epiphyte and Phorophyte Community in Plant Formations

In the immediate study areas, a total of 1090 individuals of 38 species of trees were counted. Of them, 26 species were occupied by at least one epiphyte plant with 265 phorophytes. Of this total, 1244 individuals from eight orchid species (*Broughtonia cubensis* (Lind.) Cogn., *Broughtonia ortgiesiana* (Rchb. f.) Dressler, *Encyclia bocourtii* Mujica and Pupulin, *Dendrophylax lindenii* (Lindl.) Bentham ex Rolfe, *Tolumnia guibertiana* (A. Rich.) Braem, *Tolumnia calochila* (Cogn.) Braem, *Prosthechea boothiana* (Lindl.) Higgins and *Pleurothallis caymanense* C. D. Adams. There was also a strong presence of individuals in the genus *Encyclia* that could not be identified to species level as they were sessile at the time of the survey. These individuals, belonging to one of the three species of the genus found in the study area (*E. bocourtii*, *Encyclia fucata* (Lindl.) Britton and Millspaugh and *Encyclia plicata* (Lindl.) Britton and Millspaugh were classified as *Encyclia* sp. Four species of *Tillandsias* and one of *Cactus* sp. were also found, bringing the total number of epiphytic species found at the site to 15.

To characterize the relationship between epiphyte and phorophyte communities, a multivariate analysis was carried out on the frequency matrix of 26 species, plus the "dead unidentified trees" category (Table 1). The matrix indicated the relationship of the phorophyte species with at least one epiphyte individual in the set of areas analysed.

Table 1. List of phorophyte species occupied by at least one epiphyte species, and its correspondence to Figure 1. DCA statistics (weight, N2 and heterogeneity, defined as the square root of the deviation on the coordinates)

Family	Ley	Species	Weight	N2	Het.
Anacardiaceae	Cd	*Comocladia dentata* Jacq.	18.00	3.06	71.33
Anacardiaceae	Mb	*Metopium browneii* (Jacq.) Urb.	3.00	1.00	0.00
Apocynaceae	Po	*Plumeria obtusa* L.	128.00	2.63	67.66
	Dt	Died trees	11.00	3.27	79.91
Arecaceae	Tr	*Thrinax radiata* Lodb. ex Schult. & Schult. f.	1.00	1.00	0.00
Bignoniaceae	Cm	*Catalpa macrocarpa* (A. Rich.) Ekman & Urb.	5.00	2.27	64.92
Bignonaceae	Ta	*Tabebuia angustata* Britton	2.00	2.00	58.28
Boraginaceae	Bv	*Bourreria virgata* (Sw.) G. Don.	6.00	3.00	86.02
Boraginaceae	Cg	*Cordia gerascanthus* L.	9.00	2.31	54.22
Celastraceae	Sf	*Schaefferia frutescens* Jacq.	6.00	2.57	65.35
Chrysobalanaceae	Ci	*Chrysobalanus icaco* L.	17.00	1.12	36.74
Crombetaceae	Tc	*Terminalia chicharronia* Wright, subsp. *neglecta* (Bissse) Alwan & Syace	2.00	2.00	74.22
Erythroxylaceae	Ea	*Erythroxylum areolatum* L.	55.00	4.32	99.37
Erythroxylaceae	Ec	*Erythroxylum confusum* Briton, in N. L. Britton & Millspaugh	1.00	1.00	0.00
Euphorbiaceae	Ar	*Adelia ricinella* L.	2.00	2.00	58.28
Euphorbiaceae	Gl	*Gymnanthes lucida* Sw.	37.00	4.70	117.62
Lamiaceae	Vd	*Vitex divaricata* Sw.	3.00	3.00	68.66
Malpighiaceae	Ss	*Stigmaphyllom sagranum* A. Juss	24.00	3.95	113.44
Moraceae	Fc	*Ficus crocata* (Miq.) Miq.	1.00	1.00	0.00
Moraceae	Ft	*Ficus trigonata* L.	1.00	1.00	0.00
Phyllanthaceae	Cn	*Chascotheca neopeltandra* (Griseb.) Urban	21.00	3.13	75.93
Picrodendraceae	Pb	*Pichrodendrom baccatum* (L.) Krug & Urb.	3.00	1.80	32.45
Rubiaceae	Gc	*Guettarda calyptrata* A. Rich	1.00	1.00	0.00
Salicaceae	Bb	*Banara brittonii* Roig	1.00	1.00	0.00
Sapotaceae	Sf	*Sideroxylon foetidissimum* Jacq.	6.00	2.57	65.35
Verbenaceae	Cs	*Citharexylum spinosum* L.	7.00	1.96	34.07
Verbenaceae	Li	*Lantana involucrata* L.	1.00	1.00	0.00

The results are given in Table 2 and in Figure 1. The length on the first coordinates axis showed a wide environmental gradient (>4 units d.s.), therefore the most suitable multivariate was the DCA. It also included a 45.4% of accumulated variance between the first tow axes.

Table 3 lists the epiphytic species found in the study plots, 11 in total, nine of which were orchids, in addition to the remaining statistical data from the DCA.

Figure 1 shows the distribution of epiphytes and phorophytes in relation to the Euclidean distance determined by the similarity on the plane generated by the first two axes of the DCA. Epiphytic species closer to each other shared locations on phorophytes. Likewise, phorophytes close together harboured similar epiphyte communities. Axes 1 and 2 determined a plane where the epiphytes were located.

The gradient represented in axis 1 consisted of the orchid species *E. bocourtii* at one end, and *D. lindenii* at the other, as there was a certain amount of segregation detected in the habitat, only co-habiting on the phorophytes *Plumeria obtusa* L. and *Erythroxylum areolatum* L. as hosts, out of the total on which *E. bocourtii* (9) and *D. lindenii* (11) were found. On the other hand, axis 2 was marked by the gradient with the orchids *P. boothiana*, *P. caymanensis* and *T. calochila*. At either end, sharing phorophytes like *Gymnanthes lucida* Sw. and *E. areolatum*; contrasting with *T. guibertiana*, at the other end of axis 2, the only species of epiphyte reported on the phorophyte *Banara brittonii* Roig. In the middle of the chart are orchids with lower habitat specificity, such as *B. cubensis,* and *Encyclia sp.,* and others like *B. ortgiesiana*, that colonize phorophytes with low specificity, such as *G. lucida.* Likewise, phorophytes hosting more varied flowering orchids were in a centripetal position, while less frequent hosts tended to be towards the ends, for example: *Cordia gerascanthus* L., *B. brittonii, Ficus crocata* (Miq.) Miq and *Thrinax radiata* Lodb. ex Schult. & Schult. f. in J. J. Roemer & Schult.

In addition, the statistics from the multivariate analysis can be used as indicators for various ecological properties (structure of the community) of the phorophyte-epiphyte system. Thus, in Tables 1 and 3 show the values pertaining to weight, N2 and tolerance of the phorophyte and epiphyte species respectively.

The weight refers to the factor of each phorophyte species (w) or each epiphytic specie (wk) that modulates the value of a species in the sample, or of a sample for each species. In phorophytes, the largest weights are for *P. obtusa* (128), due to the high frequency of *B. cubensis*, with *E. areolatum* and *G. lucida* farther behind.

Table 2. Detrended Correspondence analysis results

Axes	**1**	**2**	**3**	**4**	**Total inertia**
Eigenvalues	0.475	0.142	0.057	0.032	1.359
Gradient length	4.602	4.823	2.784	4.103	
Accumulated percentage of variance in the orchid species	34.9	45.4	49.6	51.9	

Table 3. List of epiphytic species and their correspondence to Figure 1. DCA statistics (weight, N2 and tolerance, defined as the square root of the deviation on the coordinates)

Epiphytic species	**Abbreviation**	**Weight**	**N2**	**Tolerance**
Broughtonia cubensis (Lindl.) Cogn.	Bc	130.00	4.18	48.35
Broughtonia ortgiesiana (Rchb. f.) Dressler	Bo	1.00	1.00	366.81
Dendrophylax lindenii (Lindl.) Bentham ex Rolfe	Dl	26.00	6.63	87.54
Encyclia bocourtii Mujica & Pupulin	Eb	49.00	5.10	81.14
Encyclia sp.	Es	119.00	4.26	43.35
Pleurothallis caymanensis C. D. Adams	Pc	3.00	3.00	230.68
Prosthechea boothiana (Lindl.) Higgins	Pb	15.00	4.79	145.14
Tillandsia sp.	T	20.00	8.70	102.07
Tolumnia calochila (Cogn.) Braem	Ts	14.00	5.16	142.06
Tolumnia guibertiana (A. Rich.) Braem	Tg	8.00	4.00	232.13

The maximum values for orchids were found with *B. cubensis* and *Encyclia* sp. In addition, N2 was the effective number of occurrences, analogous to the Hill diversity measurement (1973), which describes an index as reciprocal, proportional abundance, equivalent to the weighted measure of abundance in the occurrences. The maximum values of N2 in phorophytes were found in *G. lucida* and *E. areolatum*. Maximum values among epiphytes were obtained by the bromeliad *Tillandsia* sp. and by *D. lindenii.*

Tolerance (tk, epiphytes) or heterogeneity (hk, phorophytes) is defined as the square root of the deviation on the value of the coordinates. With the phorophyte species, calculated from the eigen values, this would be the equivalent to a variability or weighted diversity (heterogeneity) measure in the compositions of the orchid flora and other epiphytes that it can host. In the case of the latter, the tolerance will be an indicator of the amplitude of habitat, understood as the amplitude in the frequency range of orchids capable of colonizing different phorophytes. The highest heterogeneity values in phorophytes were found in *G. lucida, Stigmaphyllon sagranum* A. Juss, and *E. aerolatum*, which had a very even epiphyte flora composition. On the other hand, the lowest heterogeneity values were shown by *Erythroxylum confusum* Britton in N.L. Britton and C.F. Millspaugh, *Lantana involucrata* L., *Ficus trigonata* L., *B. brittonii, T. radiata, F. crocata, Guettarda calyptrata* A. Rich and *Metopium browneii* (Jacq.) Urb.

Concerning the orchid flora, *B. ortgiesiana* had greater tolerance value, as it colonized *G. lucida* exclusively, followed by *T. guibertiana* and *P. caymanensis*, which were the species with least specificity of habitat of those analysed (weighted by frequency of occurrence). The minimum values were obtained by the species of greatest weight, such as *B. cubensis* and *Encyclia* sp. Although they were found on a large number of phorophytes, these orchids were much more frequent on specific species and were hosted by the more generalist species.

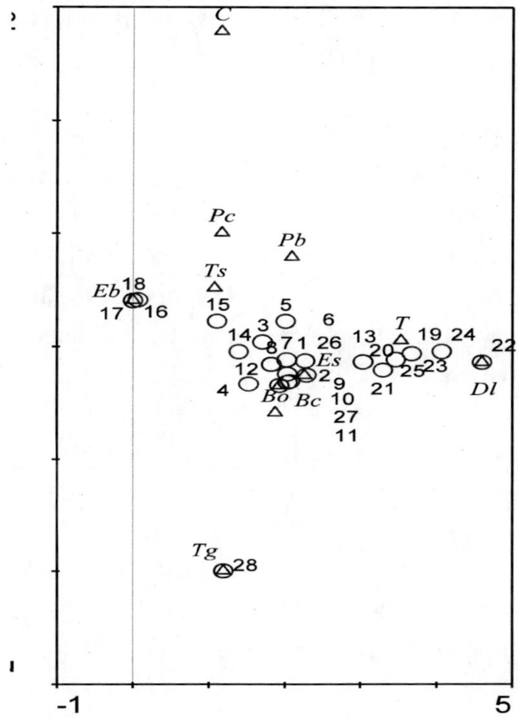

Figure 1. Representation of the DCA obtained from the frequency of phorophytes presenting orchid species and other epiphytes (abbreviations of orchid names are explained in Table 3).

CONCLUSION

Trapnell and Hamrick (2006) thought that studies on the specificity of the relationship between epiphytes and phorophytes were scarce but showed that it was highly variable. Some studies found low specificity among the epiphytes and the phorophytes they inhabited (Johansson 1974; Sanford 1974; Todzia 1986; Ackerman et al. 1989: Zimmerman and Olmsted 1992; Ackerman et al. 1996), while others found a strong association (Went 1940; Frei 1973; Ter Steege and Cornelissen 1989; Merwin et al. 2003). Bergstrom and Carter (2008) found that although occurring on 8 different hardwood tree species in Dudley's Hammock,

Epidendrum magnoliae Muhl. had a very strong preference for *Magnolia grandiflora* L. as a host and a moderately strong preference for *Quercus virginiana* Mill.

Martínez-Meléndez (2008) showed that epiphytes form generalist stands but have a weak preference for certain hosts. In their study, less than half of the epiphytic species (five) demonstrated a greater preference for at least four hosts, which suggests a low preference, which again indicates the rarity of host specificity (Benzing 1990) and the tendency of epiphytes to inhabit more than one species of tree. Their data showed that there was no host on which epiphytes concentrated their preferences, but there were some tree species that were unsuitable hosts for most epiphytes. Benavides et al. (2011) found that the large majority of epiphytes occurred in low densities on many different phorophyte species.

Nurfalidah (2015) found that some epiphytic orchid species occurred on a single phorophyte species. Other epiphytic orchid species occurred on multiple phorophyte species. Rosa-Manzano (2014) also reported that most epiphytic orchids in tropical dry forests of Yucatan, Mexico, occurred on a single tree species.

Host size could be important in determining the host–epiphyte association because size could determine the susceptibility of a host to be found by epiphyte seeds (Vergara et al. 2010). They found that forest host composition could limit the abundance of epiphytes. In the tropical dry forest studied, only 7% of the host species and 16% of the host individuals were preferred hosts for epiphytes. They recorded 72% of the epiphyte stands on these hosts, indicating a clear epiphyte preference.

In our study, the relationship between the epiphyte community and phorophyte species is generally asymmetric, so that there are generalist epiphyte species with little specificity of habitat or low requirements, and phorophytes that are also generalist and able to host several species of orchids and other epiphytes.

We also detected a very strong presence of some orchid species on individuals of certain phorophyte species, whereas in others nearby of the same species and aspect, not a single example was found. Migenis and

Ackerman (1993) in their studies on epiphytic orchids in Bisley Watersheds, Puerto Rico, did not find any specificity among the orchid species in the area and the phorophyte species. They also noted that their data suggested a preference for certain phorophyte species, not a relationship between orchid and phorophyte species. Tremblay (1998) found that *Lepanthes caritensis* Tremblay and Ackerman in Puerto Rico only colonized the *Micropholis guyanensis* A. D. C. phorophytes. Recognizing that this was a rare phenomenon in epiphytic orchids, Pierre also noted that only 7% of this species of phorophyte were colonized, although it was the most common one in the study area. Otero (2007) found that individuals of *Psychilis monensis* Sauleda, were over-represented on some phorophytes and under-represented or absent on others. Similar to its presumed sister species (*Psychilis krugii* (Bello) Sauleda), there was no relationship between phorophyte preference (Ackerman et al. 1989).

We agree with these other authors in that there could be several reasons why this phenomenon occurs. It may be because the seeds did not manage to colonize these potential phorophytes which, judging by the distribution in the areas studies seems improbable, as it would be almost impossible for the seeds to colonize phorophytes of the same species and size intermittently.

From the beginning of the 20th Century (Bernard 1909) it has been known that orchid seeds depend on specific fungi to germinate. It was even found that development of the interaction with mycorrhizae was, and still is, a crucial event in the evolution of the Orchidaceae family (Benzing 1981; Rasmussen 1995).

Whigham (2002) in his studies on the interaction of wood decomposition, mycorrhizae and seeds in species of terrestrial orchids, show that the presence of mycorrhizae was more important than the type of substrate. At the same time, they found that many species of terrestrial orchids are generalist, while others are specialist when interacting with certain mycorrhizae. Bayman (2002) notes that few studies have delved further into this aspect of epiphytic orchids, as the tiny size of the seeds makes experimenting *in situ* difficult.

Tremblay (1998) suggested that the association between orchids and some moss species plays some role in the specificity of orchids for certain species, and sizes of phorophytes could be due to the accumulation of those mosses in the species with which they have adapted to colonize.

Our study suggests that we have detected the same phenomenon. Everything seems to indicate that some orchid species colonize those phorophyte, on which the specific mycorrhizae they require for seed germination are likely present, without differentiating among species. The reason for certain phorophyte species being more frequent in this relationship may be, first and foremost, their dominance in the studied area. Secondly, they may have a stronger association with the mycorrhizae, leading to a larger presence of individuals of specific species of orchid on these. At the same time, a given orchid species occupying certain places could be ecologically conditioned by such as light, moisture, and the presence of mycorrhizae. Therefore, these species only colonize places providing the minimum requirements needed from these three factors. New studies aimed at documenting the different types of mycorrhizal fungi that may be present on the bark of these phorophytes is a logical next step towards furthering our understanding of epiphytic orchid distribution in Guanahacabies NP and beyond.

REFERENCES

Ackerman, James D. "Coping with the Epiphytic Existence: Pollination Strategies." *Selbyana* 9, no. 1 (1986): 52–60.

Ackerman, James D. "Limitations to Sexual Reproduction in *Encyclia krugii* (Orchidaceae)," Systematic Botany 14, no. 1 (1989): 101-109.

Ackerman, J. D., A. Sabat, and J. K. Zimmerman. "Seedling Establishment in an Epiphytic Orchid: An Experimental Study of Seed Limitation." *Oecologia* 106, no. 2 (1996): 192–198.

Arditti, J. *Fundamentals of Orchid Biology*. New York, John Wiley and Sons, 1992.

Bayman, Paul, Eduardo J. Gonzalez, Jose J. Fumero, and Raymond L. Tremblay. "Are Fungi Necessary? How Fungicides Affect Growth and Survival of the Orchid *Lepanthes rupestris* in the Field." *Journal of Ecology* 90, no. 6 (2002): 1002–1008.

Benavides, A. M., A. Vasco, A. J. Duque, and J. F. Duivenvoorden. "Association of Vascular Epiphytes with Landscape Units and Phorophytes in Humid Lowland Forests of Colombian Amazonia." *Journal of Tropical Ecology* 27, no. 03 (2011): 223–237.

Bennett, Bradley C. "Spatial Distribution of *Catopsis* and *Guzmania* (Bromeliaceae) in Southern Florida." *Bulletin of the Torrey Botanical Club* 114, no. 3 (1987): 265–271.

Benzing, David H. "Alternative interpretations for the evidence that certain orchids and bromeliads act as shoot parasites." *Selbyana*, 5, no 2 (1979): 135–144.

Benzing, D. H. "Why is Orchidaceae so large, its seeds so small and its seedlings mycotrophic?" *Selbyana* 5, no. 3/4 (1981): 241–242.

Benzing, David H. *Vascular Epiphytes: General Biology and Related Biota*. Cambridge University Press, 1990.

Bergstrom, Bradley J., and Richard Carter. "Host-tree Selection by an Epiphytic Orchid, *Epidendrum magnoliae* Muhl. (Green Fly Orchid), in an Inland Hardwood Hammock in Georgia." *Southestern Naturalist*, 7, no. 4 (2008): 571–580.

Bernal, Rocío, Teresa Valverde, and Laura Hernández-Rosas. "Habitat preference of the epiphyte *Tillandsia recurvata* (Bromeliaceae) in a semi-desert environment in Central Mexico." *Canadian Journal of Botany* 83, no 10 (2005): 1238–1247.

Bernard, Noël. "L'evolution dans la symbiose, les orchidées et leure champignons commensaux." *Annales des Sciences Naturelles Botaniques* 9 (1909): 1–196. ["The evolution in symbiosis, orchids and their commensal fungi." *Annals of Natural Botanical Sciences*]

Catling, Paul M., Vivian R. Brownell, and L. P. Lefkovitch. "Epiphytic orchids in a Belizean grapefruit orchard: distribution, colonization and association." *Lindleyana* 1, no. 3 (1986): 499–515.

Catling, Paul M., and Leonard P. Lefcovitch. "Associations of vascular epiphytes in a Guatemalan cloud forest." *Biotropica* 21, no. 1 (1989): 35–40.

Clements, M. A. 1987. "Orchid-fungus-host associations of epiphytic orchids." In *Proceedings of the 12th World Orchid Conference*, Tokyo, 1987, 80-83.

Ferro Jorge, Freddy Delgado and A. B. Martínez. 1995. "Mapa de Vegetación actual de la Reserva de la Biosfera Península de Guanahacabibes, Pinar del Río, Cuba." In *Memorias del II Simposio Internacional HUMEDALES'94*, Ciénaga de Zapata, Matanzas, 1995, 130-132. Ciénaga de Zapata National Park, Cuba. [Current Vegetation Map of the Guanahacabibes Peninsula Biosphere Reserve, Pinar del Río, Cuba. In *Proceedings of the II HUMEDALES'94 International Symposium*]

Ferro, Jorge. *Efectos del aprovechamiento forestal sobre la estructura y dinámica de la comunidad de epífitas vasculares del bosque semideciduo notófilo de la Península de Guanahacabibes*. PhD diss., University of Pinar del Río. Cuba, 2004. [*Effects of forest use on the structure and dynamics of the community of vascular epiphytes of the semi-deciduous notophilic forest of the Guanahacabibes Peninsula*]

Frei, S. J. K. "Orchid Ecology in a Cloud Forest in the Mountains of Oaxaca, Mexico." *American Orchid Society Bulletin* 42 (1973): 307–314.

Johansson, Dick. "Ecology of Vascular Epiphytes in West African Rain Forest." *Acta Phytogeographyca Suecica* 59 (1974): 1–159.

Krömer, Thorsten, and Robbert S. Gradstein. "Species richness of vascular epiphytes in two primary forests and fallows in the Bolivian Andes." *Selbyana* 24, no. 2 (2003): 190–195.

Krömer, Thorsten, Michael Kessler, and S. Robbert Gradstein. "Vertical Stratification of Vascular Epiphytes in Submontane and Montane Forest of the Bolivian Andes: The Importance of the Understory." *Plant Ecology* 189, no. 2 (2007): 261–278.

Lepš, Jan, and Petr Šmilauer. *Multivariate Analysis of Ecological Data Using CANOCO*. Cambridge University Press, 2003.

Martínez-Meléndez, Nayely, Miguel A. Pérez-Farrera, and Alejandro Flores-Palacios. Estratificación vertical y preferencia de hospedero de las epífitas vasculares de un bosque nublado de Chiapas, México. *Revista de Biología Tropical* 56, no. 4 (2008): 2069-2086. [Vertical stratification and host preference of vascular epiphytes from a cloud forest in Chiapas, Mexico. *Journal of Tropical Biology*]

Merwin, Mark C., Steve A. Rentmeester, and Nalini M. Nadkarni. "The Influence of Host Tree Species on the Distribution of Epiphytic Bromeliads in Experimental Monospecific Plantations, La Selva, Costa Rica." *Biotropica* 35, no. 1 (2003): 37–47.

Migenis, Luis E., and James D. Ackerman. "Orchid-Phorophyte Relationships in a Forest Watershed in Puerto Rico." *Journal of Tropical Ecology* 9, no. 2 (1993): 231–240.

Nurfalidah, Siti. "Diversity of Epiphytic Orchids and Host Trees (Phorophytes) in Secondary Forest of Coban Trisula, Malang Regency, East Java, Indonesia." *Biotropia* 22, no. 2 (2015): 120–128.

Otero, J. Tupac, Susan Aragón, and James D. Ackerman. "Site Variation in Spatial Aggregation and Phorophyte Preference in *Psychilis monensis* (Orchidaceae)." *Biotropica* 39, no. 2 (2007): 227–231.

Pérez, J. *Estudio de la distribución de epífitas sobre especies arbóreas representativas en bosques de palma*. MS Thesis. University of Puerto Rico, Río Piedras, 1982. [*Study of the distribution of epiphytes on representative tree species in palm forests*]

Rasmussen, Hanne N. *Terrestrial Orchids: From Seed to Mycotrophic Plant*. Cambridge University Press, 1995.

Rosa-Manzano, Edilia de la, José Luis Andrade, Gerhard Zotz, and Casandra Reyes-García. "Epiphytic Orchids in Tropical Dry Forests of Yucatan, Mexico – Species Occurrence, Abundance and Correlations with Host Tree Characteristics and Environmental Conditions." *Flora* 209, no. 2 (2014): 100–109.

Sanford, W. W. "The Ecology of Orchids." In *The orchids: Scientific studies*, edited by Carl L. Withner, 1–100, New York: John Wiley and Sons, 1974.

Steege, H. ter, and J. H. C. Cornelissen. "Distribution and Ecology of Vascular Epiphytes in Lowland Rain Forest of Guyana." *Biotropica* 21, no. 4 (1989): 331–339.

Ter Braak, C. J. F. "5 Ordination." In *Data Analysis in Community and Landscape Ecology*, edited by R. H. G. Jongman, C. J. F Braak, and O. F. R. Van Tongeren, 91–173. Cambridge University Press, 1995.

Ter Braak, C. J. F., and P. Šmilauer. *CANOCO reference manual and CanoDraw for Windows user's guide: software for canonical community ordination (version 4.5).* Ithaca, New York, USA: Microcomputer Power, 2002.

Todzia, Carol. "Growth Habits, Host Tree Species, and Density of Hemiepiphytes on Barro Colorado Island, Panama." *Biotropica* 18, no. 1 (1986): 22–27.

Trapnell, Dorset W., and J. L. Hamrick. "Variety of Phorophyte Species Colonized by the Neotropical Epiphyte, *Laelia rubescens* (Orchidaceae)." *Selbyana* 27, no. 1 (2006): 60–64.

Tremblay, Raymond L., Jess K. Zimmerman, Ligia Lebrón, Paul Bayman, Inés Sastre, Franklin Axelrod, and Janice Alers-García. "Host Specificity and Low Reproductive Success in the Rare Endemic Puerto Rican Orchid Lepanthes Caritensis." *Biological Conservation* 85, no. 3 (1998): 297–304.

Vergara-Torres, Carmen Agglael, Mary Carmen Pacheco-Álvarez, and Alejandro Flores-Palacios. "Host Preference and Host Limitation of Vascular Epiphytes in a Tropical Dry Forest of Central Mexico." *Journal of Tropical Ecology* 26, no. 6 (2010): 563–570.

Went, F. W. "Soziologie der Epiphyten eines tropishen Urwaldes." *Annales du Jardin botanique de Buitenzorg* 50 (1940): 1–98. [Sociology of the epiphytes of a tropical jungle. *Annals of the Buitenzorg Botanical Garden*]

Whigham, Dennis F., Jay O'Neill, Melissa McCormick, Cynthia Smith, Hanne Rasmussen, Bruce Caldwell, and Tim. Daniell. "Interactions between decomposing wood, mycorrhizas, and terrestrial orchids seeds and protocorms." In *Trends and fluctuations and underlying mechanisms in terrestrial orchid populations*, edited by P.

Kindlmann P., J. H. Willems, D. F. Whigham, 117–131, Leiden, The Netherlands: Backhuys Publishers, 2002.

Zimmerman, Jess K., and Ingrid C. Olmsted. "Host Tree Utilization by Vascular Epiphytes in a Seasonally Inundated Forest (Tintal) in Mexico." *Biotropica* 24, no. 3 (1992): 402–407.

Zotz, G., and P. Hietz. "The Physiological Ecology of Vascular Epiphytes: Current Knowledge, Open Questions." *Journal of Experimental Botany* 52, no. 364 (2001): 2067–2078.

Zotz, G. and G. Schmidt. "Population decline in the epiphytic orchid *Aspasia principissa*." *Biological Conservation* 129, no. 1 (2006): 82–90.

Reviewed by:

Lawrence W. Zettler
Illinois College, IL, USA.

In: Orchidaceae
Editor: Vladan Djordjević

ISBN: 978-1-68507-189-9

Chapter 6

ECOLOGY AND POPULATION STRUCTURE OF *ENCYCLIA BOCOURTII* PUPULIN AND MÚJICA (ORCHIDACEAE), A SPECIES ENDEMIC TO THE GUANAHACABIBES NATIONAL PARK, PINAR DEL RÍO PROVINCE, CUBA

Elaine González Hernández[1,*], Andreu Bonet[2], Ernesto Mújica Benítez[1], Alejandro Camejo[3] and Josep Raventós[2]

[1]Jardín Botánico Orquideario Soroa,
Universidad de Artemisa, Candelaria, Cuba
[2]Departamento de Ecología. Universidad de Alicante,
San Vincent del Raspeig, Alicante, Spain
[3]Parque Nacional Guanahacabibes. Sandino, Pinar del Río, Cuba

* Corresponding Author's E-mail: egh@uart.edu.cu.

ABSTRACT

We analyzed some structural characteristics of a population of *Encyclia bocourtii*, an epiphytic orchid endemic to Cabo San Antonio on the Guanahacabibes Peninsula of western Cuba. This species displayed an aggregate distribution pattern with high co-specific competition along the five transects surveyed. It was selectively found on only three species of phorophytes, *Chrysobalanus icaco*, *Metopium brownei* and *Thrinax radiata*, predominately on individuals with rugose or semi-rugose bark. In Cabo San Antonio, *E. bocourtii* has a metapopulational structure with highest density noted in the area of coastal vegetation. A hypothesis as to the quality of sites is discussed in relation to the possible presence of mycorrhizae.

Keywords: Orchidaceae, *Encyclia bocourtii*, structure, ecology, Cuba

INTRODUCTION

As one of the two most diverse groups of angiosperms along with Asteraceae (WCSP 2021), the family Orchidaceae with ca. 27,000 species has received considerable attention due to its enormous economic and social value compared with other species (Kindlmann et al. 2002). This family is probably the most highly evolved of all plants and constitutes a very important component of global biodiversity. In relation to other plant families, orchids are subject to a higher risk of extinction, due to anthropogenic factors and/or natural causes (Hutchings 1989). Studies carried out in recent years show that many species are now vulnerable to extinction (Swartz and Dixon 2009; Warseno and Bagus 2013) because little is known about their complex life cycles and their populations remain unstudied. Most (ca. three fourths) of orchid species are epiphytic, yet the majority of all population studies involve temperate terrestrial species (Kull 2002; Kindlmann et al. 2002), and only few of them treat epiphytic or lithophytic orchids (Zotz and Tyree 1996; Zotz 1998, 1999, 2000; Tremblay 1997; Tremblay and Hutchings 2003).

In Cuba, orchids are considered to be the best-represented plant family among the archipelago's vascular flora (Díaz 1996), 30.5% of which are endemic (Mújica and González 2015). One example is *Encyclia bocourtii*, endemic to the Guanahacabibes Peninsula National Park in Pinar del Río Province, where it is found primarily in coastal sandy vegetation. This species is also found in xeromorphic coastal vegetation and in semideciduous forests, where it is less abundant. *Encyclia bocourtii* is also an epiphytic species found on different phorophytes that are either isolated or in aggregation with other individuals of the same species.

The purpose of our study was to evaluate some aspects of the structure and population ecology of this rare endemic species, and to answer the following questions:

- Does *E. bocourtii* exist as a generalist or specialist species on the phorophytes it colonizes?
- Does this species show some preference for the bark type associated with phorophytes?
- What is the pattern of spatial distribution of this species?
- Is intraspecific competition an important interaction in this species?
- What is the pattern of spatial distribution among the different life classes in *E. bocourtii*?
- How does the structure of life classes behave under isolated versus aggregated conditions?

METHODS

Study area - The Guanahacabibes Peninsula National Park (G.N.P.) is located in the westernmost part of the island of Cuba, specifically in the municipality of Sandino, Pinar del Río Province. The GNP is bordered in the north by the Gulf of Mexico, in the south by the Caribbean Sea, in the east by the La Fé-Manuel Lazo demarcation and in the west by the Strait of Yucatan. It is an elongated territory, narrow and flat, comprising ca. 101,500 hectares, of which 28,000 belong to the terrestrial area of the Park. In regard to its climate, the peninsula differs little from the rest of the territory of western Cuba, although its great length and narrow width render the peninsula more influenced by the sea surrounding it and the Gulf of Mexico's currents. The average annual precipitation is 1334 mm (Lopetegui et al. 1999), which is not evenly distributed, the averages in the southern and northern parts of the peninsula being 1000-1200 mm and 1400-1600 mm, respectively. The relative humidity remains almost invariable, with an average of 81% (Lopetegui et al. 1999). The minimum and maximum annual averages of temperature are 21.5 and 29.2 ºC, respectively. The prevailing winds are from the northeast throughout the whole year. The drought period extends from November until April. In the G.N.P., 10 different vegetation formations are found that support 14 endemic species. The semideciduous forest is the predominant one over all of its extension, mainly on karstic ground in the central part. A total of 750 plant species have been identified on the peninsula, of which 500 are widespread, 231 are endemic and 19 are endemic to the peninsula (Urquiola pers. com.). Most of the ground consists of compact exposed karstic limestone.

This study utilized data collected during 15 consecutive years (2006-2021) using transects. The transects were the same as those perpendicular to the coastline that were used to establish the population's limits. This species is located almost exclusively on sandy coastal vegetation in a narrow strip in the area of Punta Cajón on the northwestern side of the peninsula. This strip rarely exceeds 40 m in width. We decided to trace five transects, one horizontal to the coastline with S-N orientation, 100 m long and 15-20 m wide. Measuring about 40 m in length and 20-25 in width, the other four were positioned perpendicular to the coastline with an E-W orientation to more accurately determine the pattern of distribution of the species at this site.

Individuals of *E. bocourtii* are typically medium-sized to large, with 1-3 leaves in each pseudobulb. Their flowers are showy, with a pleasant light fragrance, medium-sized, with yellowish sepals and petals and a white labellum with brown stripes. The flowering period occurs mostly in June, extending until July (Pupulin and Mújica 2005).

All of the known phorophytes were marked, and we identified each to the species level. Data on the height and DBH of phorophytes, and on the height, DBH and orientation of each individual of *E. bocourtii* were collected to determine spatial distribution preferences.

For delimitation of each life class, the biological method was chosen, using the size approach (García Begoña 2000). Accordingly, the population was divided into four classes: seedlings (Class 1), juveniles (Class 2), reproductive adults I (Class 3) and reproductive adults II (Class 4). To our knowledge, only a few studies have been published involving the population structure of tropical orchids. We therefore decided to divide individuals of the population into two main groups: reproductive and non-reproductive individuals, and each group into two classes, C-3 and C-4 (reproductive) and C-1 and C-2 (non-reproductive). Leaf area was the variable used to define the class to which each individual belongs because a close relationship exists between leaf area and size. For this purpose, the length and width of all leaves in every individual were measured to calculate an average for each one.

To separate the C-3 class from the C-4 class (reproductive group), leaf area was calculated for all individuals in this group. The average was 22.5 cm^2, and we therefore decided that all individuals with leaf area > 22.5 cm^2 would be categorized as C-4, while those that had a smaller leaf area would be relegated to the C-3 class. Subdivision of the non-reproductive group was more difficult to ascertain because the seedlings were very small in comparison with the rest of the classes. Consequently, we relied on our direct observations. Comparing leaf areas and sizes in seedlings *in situ*, we found that they did not exceed 1.8 cm^2. Thus, all individuals with less than this leaf area would belong to the C-1 class, while those that presented a larger area would belong to the class of juveniles or C-2.

To study the effect of competition, we marked and identified all individuals with a radius smaller than 20 cm among each of the *E.*

bocourtii individuals present. In so doing, we postulated that in the case of a specimen with a larger radius, competition was not significant based on the typical size of individuals comprising this species. In like manner, individuals that were located < 20 cm apart from each other were deemed aggregated (in competition for space), whereas the others, located > 20 cm apart from each other, were considered isolated.

Statistics

Non-parametric statistical tests were applied to detect significant differences among some of the analyzed variables (Sokal and Rohlf 1995).

RESULTS

Analysis of Structure of the Occupied Phorophyte Community

The tree community of the coastal sandy vegetation is not species-rich. Only a few tree species were colonized by *E. bocourtii*. For analyses of the interrelation of this orchid species with the tree community, we decided to include seven tree species and the "dead tree" category due their importance in the ecology and dynamics of orchids (Table 1). This area was impacted by two hurricanes of great intensity, hurricane Ivan in 2004 and hurricane Wilma in 2005, that caused catastrophic losses to the sandy coastal vegetation, including the orchid population.

During the 15 years of study, 1465 individuals of this orchid species were documented, 455 of which remain intact at the time of this writing. The cause of that drastic decrease was the high number of individuals inhabiting dead trees or branches and fallen trunks. When individuals of *E. bocourtii* fall to the ground, they eventually die because they are not able to adapt to a terrestrial environment. Also during these 15 years of study, we detected a total of 272 individuals that had fallen to the ground (18.6% of the whole population) and only 13 of these currently survive. Another 14 individuals inhabited dead trees or trunks and fallen branches at the time of this writing.

Table 1 shows the results of investigating the interrelation of *E. bocourtii* with the tree community in sandy coastal vegetation. Four species were considered to be abundant in this ecosystem, viz., *Metopium brownie* (Jacq.) Urb., *Chrysobalanus icaco* L., *Thrinax radiata* Lodd. ex Schult. and Schult. and *Coccoloba uvifera* L. *Metopium brownei* had a higher orchid colonization percentage (28.5%). However, *C. icaco* was the primary phorophyte for *E. bocourtii*, given that 56.8% of all individuals that have existed or exist at the moment in the population inhabit that species of tree, with an average density of 23.7 individuals of this orchid per phorophyte. It is very interesting that *C. uvifera*, which is considered to be very common in this ecosystem, was not occupied on any occasion. We speculate that the cause may lie in the smooth bark surface, which does not provide a suitable anchor point for wind-blown orchid seeds. It is also possible that the bark substrate lacks suitable mycorrhizal fungi that would facilitate orchid seed germination leading to the orchid's colonization.

The distribution of *E. bocourtii* on phorophytes of different species (Figure 1) shows greater equality in the species *M. brownei* (Mb) and *T. radiata* (Tr). Even so, all species showed highly significant differences in the distribution of *E. bocourtii* on phorophytes. We also found considerable differences in the distribution on *C. icaco* (Ci) and the category of dead trees (Dt). For example, *C. icaco* turned out to be the species with the highest number of *E. bocourtii* individuals, but phorophytes of this species were found to harbor only one orchid individual, while those of other species were colonized by more than 100 individuals, with one even being occupied by as many as 225 individuals. In general, the distribution of *E. bocourtii* on its phorophytes showed significant differences ($p < 0.002$). Thus, it is conceivable that a very close relationship exists between phorophytes and the establishment of orchids, perhaps linked to the presence of mycorrhizal fungi.

Analysis of Bark Type

For this analysis, the category "dead trees" was not included because that would require us to know the identity of the original phorophyte species in order to establish the bark type. Figure 2 shows that 39 phorophytes have rough bark (R), as in the species *M. brownei*; 35 have semi-rough bark (SR), as in the phorophytes of *C. icaco*; and only 18 have smooth bark (S), as in *T. radiata*. A total of 74 phorophytes (80.4%) harboring *E. bocourtii* had rough or semi-rough bark, thus demonstrating the orchid's preference for these bark types.

Table 1. Interrelation between *E. bocourtii* and its phorophyte community

No.	Species of trees	Number of individuals	Occupied	%	Number of individuals of *E. bocourtii*	% of the total	Median density +d.s.
1	*Metopium brownei* (Jacq.) Urb.	137	39	28.5	290	19.8	7.44 + 11
2	*Chrysobalanus icaco* L.	174	35	20.1	832	56.8	23.77 + 43.95
3	Dead trees	70	28	40	293	20	10.46 + 10.93
4	*Thrinax radiata* Lodd. ex Schult. and Schult.	196	18	9.2	50	3.4	2.78 + 1.87
5	*Coccoloba uvifera* L.	159	0	0	0		
6	*Erythrina cubensis* C. Wright	5	0	0	0		
7	*Chloroleucon mangense* (Jacq.) Britton and Rose	1	0	0	0		
8	*Cordia nítida* Vahl.	1	0	0	0		
	Total	743	120	16.2	1465		

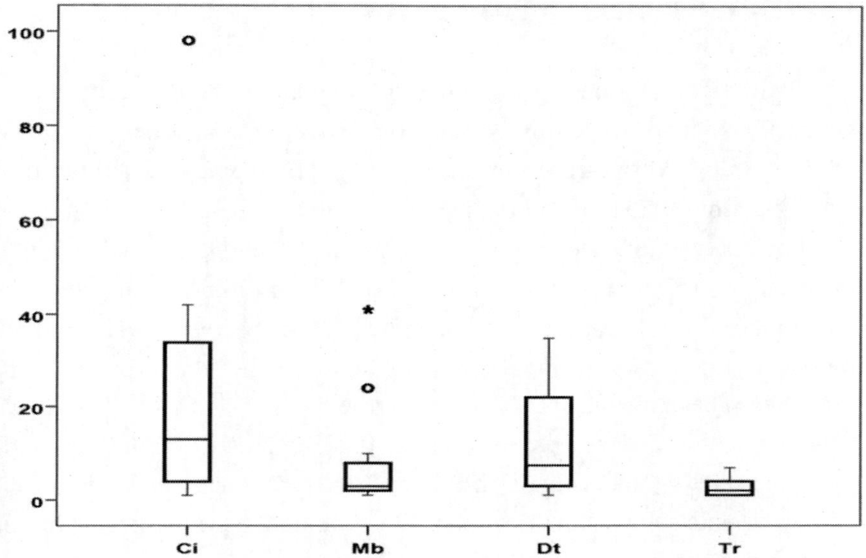

Figure 1. Distribution of *E. bocourtii* on its phorophyte species.

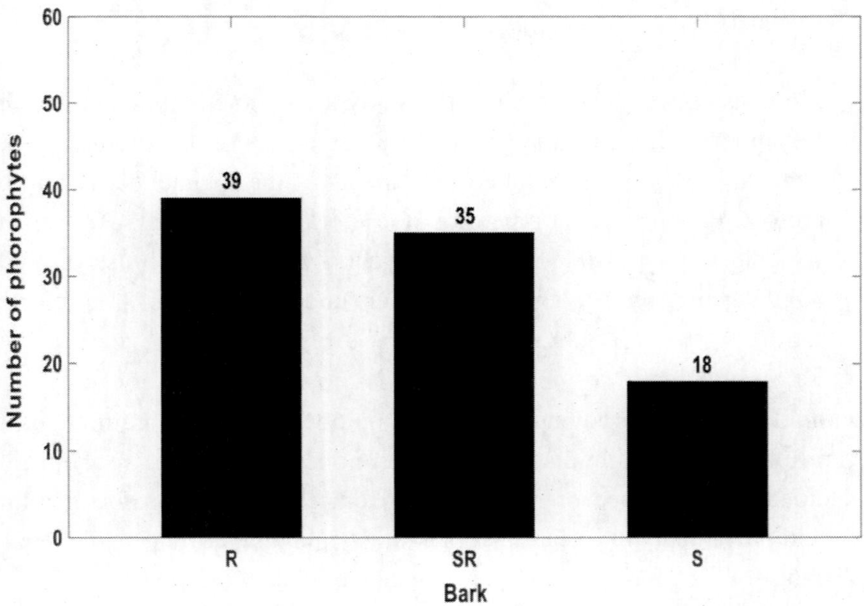

Figure 2. Preference of *E. bocourtii* for bark types.

Spatial Distribution on Phorophytes

The height of trees in coastal sand vegetation of the study area is generally low, so to find individuals of *E. bocourtii* at a height of more than 1.5 m is a very rare occurrence. The height average turned out to be very low, only 0.59 cm, and the DBH average was 8.92 cm.

For analysis of orientation, we selected the eight cardinal points of the compass (N, S, E, W, NE, NW, SE, SW) and a ninth category, viz., IND (indefinite), because some individuals of *E. bocourtii* were found to be occupying sites with indefinite orientation, either above or below the branches. These results are shown in Figure 3.

Significant differences exist ($p < 0.003$), probably caused by the small number of individuals in NE and IND. A clear preference for a certain orientation was not detected. We conclude that this species prefers a very open environment with direct solar irradiation.

Competition

In coastal sand vegetation of the study area, the epiphyte community is very poor and exemplified by the presence of only one orchid species (*E. bocourtii*) and some isolated individuals of the Bromeliaceae family. Therefore, interspecific competition does not appear to exist. To evaluate if possible intraspecific competition occurs, we decided to use only the 2020 data in order to have a clearer understanding of the current situation.

The population of *E. bocourtii* is now composed of only 455 individuals, and surveying this small number of plants required us to carefully define spatial distances. As mentioned previously, we arbitrarily decided on a distance of 20 cm apart from each other for documenting orchid individuals in isolation versus those in aggregation (i.e., those at a distance of 20 cm or greater apart from each other = isolated).

With allowance for size of the plants, we categorized those that were at 20 cm or more apart from another individual as isolated, while those that were located less than 20 cm one from one another were declared to be aggregated or in competition for space. The results show a prevalence of aggregation: only 97 individuals occurred in isolated conditions (21.3%), whereas 358 were aggregated (78.7%). This behavior argues in support of the premise that intraspecific competition is very strong.

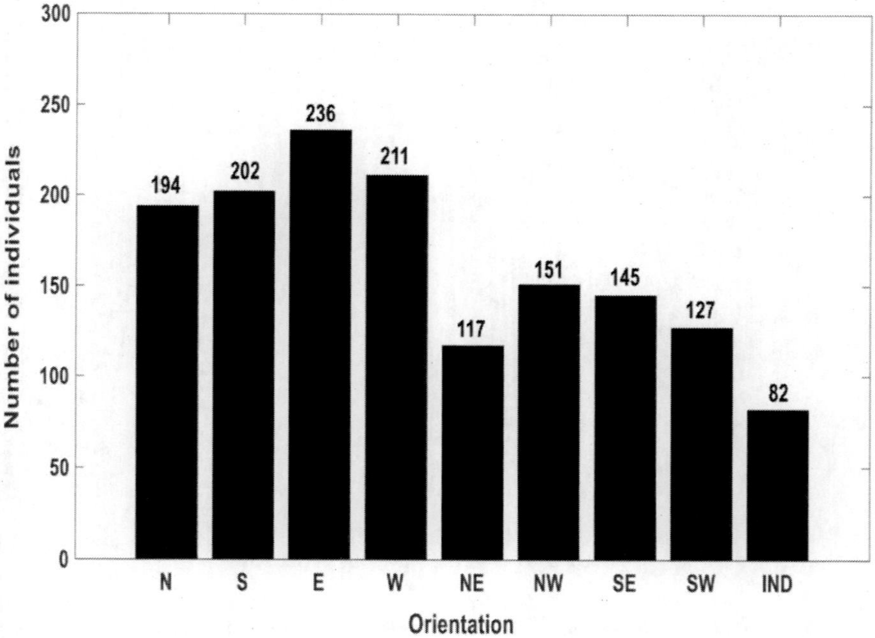

Figure 3. Orientation of sites where *E. bocourtii* individuals are located.

However, when we analyzed the current population structure in terms of life classes, we observed significant differences of structure in isolated and aggregated conditions.

Figure 4 shows that the aggregated population structure was completely asymmetric for a high number of individuals in the C-1 class (seedlings), while the structure in isolated conditions was more symmetric. The high concentration of aggregated seedlings provides evidence indicating that the seeds of this species have a high viability, but most of them germinate in small spaces. This fact implies that competition for a suitable substrate is very strong, and mortality is also very high. In the other life classes, significant differences were not detected, since even under isolated conditions the C-4 class had a higher number of individuals.

Figure 4. Life class structure of *E. bocourtii* populations in two conditions, aggregated and isolated.

CONCLUSION

The Guanahacabibes National Park remains a fragile ecosystem that harbors a large number of plant and animal species. The family Orchidaceae is broadly represented on the peninsula. The last published study indicated the presence of 28 orchid species (Ferro et al. 1995), including six endemic species, two of which were endemic to this national park (Mújica, pers. comm.). Although several studies describing the habitat of some of these species have been carried out (Ferro, unpublished), they did not include the objective of ascertaining the complex dynamics of the different life classes that compose the orchid populations which exist on the peninsula, in many cases in different ecosystems.

Encyclia bocourtii occupies coastal sand vegetation, mostly in Punta Cajón, Cabo San Antonio (G.N.P.). Previous studies in this area with other orchid species showed a very close relationship between orchid species and the phorophyte community. González (2007) found that *Broughtonia cubensis* (Lindl.) Cogn. colonized 14 different species of phorophytes, but two of them, *Plumeria obtusa* L. and *Erythroxylom aerolatum* L., represented 60% of all phorophytes with 60.2% of all the individuals of *B. cubensis*. This result demonstrates a strong preference for these two phorophyte species.

Another study carried out with a second orchid species in Cabo San Antonio, *Dendrophylax lindenii* (Lindl.) Bentham ex Rolfe (Mujica et al. 2018), showed that this species colonized 17 different phorophytes. Three of them (*Comocladia dentata* Jacq., *Diospyrus crassinervis* Krug. et Urban and *Cedrela odorata* L.) represented 39.8% of all available phorophytes and 41% harbored individuals of *D. lindenii*. *Encyclia bocourtii* was found to have a high preference for two phorophyte species (*M. brownei* and *C. icaco*). However, *C. icaco* is the most important phorophyte for this species because more than half of all *E. bocourtii* individuals colonize this species.

Bark has been a topic of discussion in connection with its use as an ecological indicator in epiphytism studies. The debates have been centered on the exact causes of the important role(s) of bark, recognizing that there are two basic elements: its physical structure (bark type) as an anchor point and its composition as a supplier of necessary chemical elements. Cuello (1998) pointed out that for epiphytes, as for any other sessile organism, characteristics of the substrate, especially during seedling establishment, are particularly important, especially the type of bark surface.

Some factors that can influence the distribution patterns and abundance of vascular epiphytes in tropical forests are linked to the architecture, behavior and characteristics of the phorophytes' bark (Lüttge 1989; Steege and Cornelissen 1989).

Our hypothesis is centered around the following premise: an orchid species does not choose a certain bark type, but rather orchid seeds colonize those spaces where mycorrhizal fungi are present. Without these fungi, the emergence of spontaneous orchid seedlings would not be possible. This hypothesis is supported by the fact that in a small area we found some trees of the same species occupied by epiphytes (phorophytes), while others at a distance of 1-2 m away were empty. Previous studies showed that *B. cubensis* prefers both rough and smooth bark, as indicated by the large number of phorophytes of the species *P. obtusa* (smooth) and *E. areolatum* (rough) (González et al. 2007). In the case of *D. lindenii*, this species has a strong association with phorophytes with rough bark (Mujica et al. 2018). Our study experimentally demonstrated *E. bocourtii*'s preference for phorophytes with rough and semi-rough bark. *Dendrophylax lindenii* also grows primarily on phorophytes with rough or semi-corrugated bark, yet this species is far less specific for phorophyte hosts compared to *E. bocourtii*. In addition, *D. lindenii* often grows in isolation on a given tree, as opposed to *E. bocourtii*, which typically grows in clusters. These observations suggest that *E. bocourtii* could be targeting specific mycorrhizal fungi that, in some way, may be connected to a narrow (select) group of phorophytes.

Studies with *B. cubensis* (unpublished data of Mujica) showed that the average height and DBH of sites occupied by individuals of this species on its phorophytes were 1.33 m and 10.9 cm, respectively. On the other hand, in the case of *D. lindenii* the height is very similar, 1.3 m, while the DBH average is 27.1 cm (Mujica 2008). As for orientation, *B. cubensis* clearly preferred N and E orientation, while *D. lindenii* preferred E orientation (Mujica 2008). Individuals of *E. bocourtii* preferred heights and DBH much lower than in those species, while a clear preference did not exist for a certain orientation.

The results on competition support the findings of González (2007) with *B. cubensis* and Mújica (2008) with *D. lindenii* in Cabo San

Antonio. The competition in *E. bocourtii*, as in those species, is intraspecific.

The life class structure (isolated versus aggregated conditions) of *E. bocourtii* showed that individuals of this species mainly exist in aggregation, which also supports the findings by González (2007) on B. cubensis. However, they appear to differ from those obtained by Mujica (2008), who reported that *D. lindenii* prefers to grow in isolation.

This study serves as the first report documenting the ecology and structure of the population of *E. bocourtii* in Cabo San Antonio, Guanahacabibes Peninsula, where it is endemic. Future studies are currently under way with this and other orchid species native to the region.

References

Cuello, A. A. *Review of sampling procedures for the study of vascular epiphytic species diversity in neotropical montane forest.* Herbario Universitario PORT, Programa de Recursos Naturales Renovables, UNELLEZ-Guanare, Venezuela, (1998): 11p.

Díaz, M. A. *Las orquídeas de Cuba*. La Habana: Editorial Científico Técnica, 1996: 46 p. [*The orchids of Cuba*. Havana: Technical Scientific Editorial]

Ferro Díaz, Jorge, F. Delgado; A. B. Martínez; et al. *Mapa de Vegetación actual de la Reserva de la Biosfera Península de Guanahacabibes, Pinar del Río, Cuba, Memorias del II Simposio Internacional HUMEDALES´94*. Editorial Academia, (1995): 130-32. [*Current Vegetation Map of the Guanahacabibes Peninsula Biosphere Reserve, Pinar del Río, Cuba, Proceedings of the II HUMEDALES'94 International Symposium.*]

García Begoña, B. A. *Biología de la Conservación de plantas amenazadas.* Organismo Autónomo Parques Nacionales. Madrid, España, (2000). [*Conservation Biology of threatened plants*.]

González, Elaine, Josep Raventós, Ernesto Mujica, and Andreu Bonet. "Estructura y ecología de la población del endemismo cubano Broughtonia cubensis (Orchidaceae), en Cabo San Antonio, Península de Guanahacabibes, provincia de Pinar del Río, Cuba." *Lankesteriana* 7, no. 3 (2007): 469-478. [Structure and ecology of the population of the Cuban endemism Broughtonia cubensis (Orchidaceae), in Cabo San Antonio, Guanahacabibes Peninsula, Pinar del Río province, Cuba.]

Hutchings, M. J. "Conservation and the British orchid flora". *Plants Today* 2, (1989): 50-58.

Kindlmann, P., J. H. Willems and D.F. Whigham. *Trends and Fluctuations and Underlying Mechanisms in Terrestrial Orchid Populations*. Leiden: Backhuys Publishers, 2002.

Kull, Tiiu. "Population dynamics of north temperate orchids." In *Orchid Biology: Review and Perspectives VIII,* edited by Tiiu Kull, and Joseph Arditti, 139–167, Kluwer Academic Publishers, 2002.

Lopetegui, C. M. *Caracterización climática y bioclimática de la Península de Guanahacabibes.* Memorias del Taller Internacional TROPICO´99. Instituto de Meteorología. Ciudad de La Habana, (1999): 47 p. [*Climatic and bioclimatic characterization of the Guanahacabibes Peninsula.*]

Lüttge, U. "Vascular Epiphytes: Setting the Scene." In Vascular Plants as Epiphytes. *Ecological Studies (Analysis and Synthesis),* vol 76, edited by U. Lüttge. Berlin, Heidelberg: Springer-Verlag, 1989.

Mujica, Ernesto. 2008. *Ecología de las orquídeas epífitas Cattleyopsis cubensis (Lindley) Cogniaux, Dendrophylax lindenii (Lindley) Bentham ex Rolfe y Encyclia bocourtii Mujica et Pupulin en el Cabo San Antonio, Península de Guanahacabibes, Cuba. Análisis espacio-temporal e implicaciones del impacto de un fenómeno atmosférico severo.* PhD Dissertation, Universidad de Pinar del Río, (2007): 330p. [*Ecology of the epiphytic orchids Cattleyopsis cubensis (Lindley) Cogniaux, Dendrophylax lindenii (Lindley) Bentham ex Rolfe and Encyclia bocourtii Mujica et Pupulin at Cabo San Antonio,*

Guanahacabibes Peninsula, Cuba. Spatio-temporal analysis and implications of the impact of a severe atmospheric phenomenon.]

Mújica, Ernesto, and Elaine González. "A New Checklist of Orchid Species from Cuba." *Lankesteriana* 15, no. 3 (2015). 219–69.

Mújica, Ernesto B., Justin J. Mably, Shannon M. Skarha, Laura L. Corey, Larry W. Richardson, Mark W. Danaher, Elaine H. González, and Lawrence W. Zettler. "A Comparision of Ghost Orchid (*Dendrophylax lindenii*) Habitats in Florida and Cuba, with Particular Reference to Seedling Recruitment and Mycorrhizal Fungi." *Botanical Journal of the Linnean Society* 186, no. 4 (2018): 572–586.

Pupulin, Franco, and Ernesto Mújica Benítez. "Another New Species of Encyclia (Orchidaceae: Laeliinae) from Cuba." *Harvard Papers in Botany* 10, no. 2 (2005): 227–30.

Sokal Robert R. and F. James Rohlf. *Biometry: The Principles and Practices Of Statistics In Biological Research* 3rd. New York: Freeman and Company, 1995.

Steege, H. ter, and J. H. C. Cornelissen. "Distribution and Ecology of Vascular Epiphytes in Lowland Rain Forest of Guyana." *Biotropica 21*, no. 4 (1989): 331–339.

Swartz, Nigel D. and Kingsley, Dixon. "Terrestrial orchid conservation in the age of extinction." *Annals of Botany 104*, no. 3 (2009): 543.546.

Tremblay, Raymond L. "Distribution and Dispersion Patterns of Individuals in Nine Species of Lepanthes (Orchidaceae)." *Biotropica 29,* no. 1 (1997): 38–45.

Tremblay, Raymond L., and Michael J. Hutchings. "Population Dynamics in Orchid Conservation: A Review of Analytical Methods Based on the Rare Species Lepanthes eltoroensis." In *Orchid Conservation*, edited by K. W. Dixon, S.P. Kell, R. L. Barrett, and P. J. Cribb, 163–183. Sabah: Natural History Publications Kota Kinabalu, 2003.

Warseno, Tri and Ida Bagus Ketut Arinasa. "Epiphytic Orchid and Host Tree Inventory in Bulu Pattanunang Protected Forest Maros South Sulawesi." *Proceedings of 4th International Conference on Global Resource Conservation & 10th Indonesian Society for Plant*

Taxonomy Congress. Brawijaya University, Indonesia, February 7-8. 2013.

WCSP. *World Checklist of Selected Plant Families.* Kew: Royal Botanic Gardens, http://apps.kew.org/wcsp/, 2021.

Zotz, Gerhard, and Melvin T. Tyree. "Water Stress in the Epiphytic Orchid, Dimerandra emarginata (G. Meyer) Hoehne." *Oecologia 107*, no. 2 (1996): 151–59.

Zotz, Gerhard. "Demography of the Epiphytic Orchid, Dimerandra emarginata," *Journal of Tropical Ecology* 14, no. 06 (1998): 725–41.

Zotz, G. "What Are Backshoots Good for? Seasonal Changes in Mineral, Carbohydrate and Water Content of Different Organs of the Epiphytic Orchid, Dimerandra emarginata." *Annals of Botany 84*, no. 6 (1999): 791–98.

Zotz, Gerhard. "Size Dependence in the Reproductive Allocation of Dimerandra emarginata, An Epiphytic Orchid." *Ecotropica 6*, no. 1 (2000): 95–98.

Reviewed by:

Lawrence W. Zettler
Department of Botany, Illinois College, IL, USA.

In: Orchidaceae
Editor: Vladan Djordjević
ISBN: 978-1-68507-189-9

Chapter 7

WHY IS *EPIPACTIS HELLEBORINE* (L.) CRANTZ SO WIDESPREAD?

***Agnieszka Rewicz*[1,*], *Wojciech Adamowski*[2] *and Marta Kolanowska*[3,4]**

[1]Department of Biogeography, Paleoecology and Nature Conservation, University of Lodz, Lodz, Poland
[2]Białowieża Geobotanical Station, University of Warsaw, Białowieża, Poland
[3]Department of Geobotany and Plant Ecology, University of Lodz, Lodz, Poland
[4]Department of Biodiversity Research, Global Change Research Institute AS CR, Brno, Czech Republic

ABSTRACT

The broad-leaved helleborine (*Epipactis helleborine* (L.) Crantz) is one of the most widely distributed orchids in the temperate zone. In its primary distribution, apart from various natural habitats, it is the most

[*] Corresponding Author's E-mail: agnieszka.rewicz@biol.uni.lodz.pl.

common orchid species occupying anthropogenic habitats, even in city centers. Moreover, it is also one of a few species spreading beyond their primary distribution. Undoubtedly, this species' success is stimulated by immense phenotypic plasticity and broad ecological amplitude. The broad-leaved helleborine is characterized by a wide spectrum of pollen vectors, and due to its self-compatibility, even a single individual could start a new population. The fruit set is usually high. *E. helleborine* produces a large number of seeds, which are anemochorous. In unfavorable conditions, individuals of this orchid can stay dormant for many years. Unlike numerous orchids whose occurrence is limited by the presence of specific mycorrhizal fungus, *E. helleborine* has a broad spectrum of fungal symbionts. In its secondary distribution, the broad-leaved helleborine was often intentionally planted in new locations, and this factor seems to be very important for its fast spread in North America.

Unfortunately, data about ecology of *E. helleborine* are scattered throughout various papers and reports, sometimes fragmentary and difficult to obtain (the so-called "grey literature"). Scientific elaboration on this topic still lacks a comprehensive review and summary of the biological and ecological aspects which may explain the scale of the broad distribution of *E. helleborine*. The aim of this article has been to gather, review, and analyze characters of this species which have allowed its spread into various habitats.

Keywords: *Epipactis helleborine,* Orchidaceae, anthropogenous habitats, species diversity, orchid seeds

INTRODUCTION

The Orchidaceae family is renowned for its mycorrhizal specificity, pollinator specialization and limited germination rates (Gravendeel et al. 2004; Cozzolino and Widmer 2005) that result in very limited geographical ranges of the family representatives (McCormick and Jacquemyn 2014). These plants are also susceptible to habitat alterations (Swarts and Dixon 2009; Fay 2018). There are, however, some exceptions of this general vulnerability of orchids.

The broad-leaved helleborine (*Epipactis helleborine* (L.) Crantz) is a terrestrial orchid naturally distributed from Europe to North Africa, southern Siberia, and the Himalayas (Hultén and Fries 1986; Delforge 2006, Kolanowska 2013, Evans and Jacquemyn 2020.). It occurs also in Japan (Suetsugu 2013). *Epipactis helleborine* was introduced to North America over 130 years ago and is now common in eastern Canada as well as on the western and eastern coasts of the USA (Zenkert 1949; Squirrell et al. 2001; Kolanowska 2013). It grows in broadleaved, mixed and coniferous (also secondary) forests, on forest edges, as well as in anthropogenic habitats (Hollingsworth and Dickson, 1997; Light and MacConaill 2006; Stešević and Jovanović 2008; Milović and Mitić 2012; Akhalkatsi et al. 2014; Rewicz et al. 2017, 2018, Süveges et al. 2019; Djordjević et al. 2020a, b).

Delforge (2006) has reported that this species is a pioneer species. The question is why? The main aim of this article has been to assemble, review, and analyze data about this phenomenon.

METHODS

The article has been prepared based on the available literature, including information published in scientific articles and data from the so-called "grey literature," such as local monographs or technical reports, as well as our research.

EPIPACTIS HELLEBORINE IN ANTHROPOGENIC HABITATS

The first report about the appearance of orchids in anthropogenic areas came from the 19th century, when those plants were observed at railway embankments in Great Britain (Owen 1879; Day 1882; Procházka and Velisek 1983). Adamowski (2006) reported that 53 species out of 300 European taxa might be encountered in disturbed habitats. Findings of Rewicz et al. (2017) show that *E. helleborine* is the most frequently occurring species in anthropogenic habitats. Research by

Rewicz el al. (2017) indicates that orchids were found in eight categories of anthropogenic habitats. *E. helleborine* occurred in all types of habitats: forests influenced by industrial emissions (Świercz 2004, 2005, 2006, 2007); industrial terrains (Misiewicz 1976; Rebele 1988); industrial waste places (Remmel 1970; Klimko et al. 2004); greenery: parks, hedges, etc. (Sinker et al. 1991; Godefroid 1995; Wittig and Wittig 2007); plantations of ecologically alien trees (Berezutsky et al. 2014); roadsides and embankments (Geisselbrecht-Taferner and Mucina 1995; Soczewka 2000; Jakubska et al. 2006; Rewicz et al. 2017, 2018); sand pits, clay pits, and quarries (Piotrowska 1966; Ćwikliński 1971; Nikolaeva and Zefirov 1971; Renz et al. 1984; Jackowiak 1986; Wika 1986; Berdowski and Spałek 1997; Sauerland 1995; Sudnik-Wójcikowska 1987; Szlachetko 1988; Dickson 1990; Stolarz 1994; Adamowski 1996, 1998; Nowak 1997; Urbisz and Urbisz 1998; Soczewka 2000; Bernacki and Błońska 2006, Kiedrzyński and Stefaniak 2011) (Figure 1). Also, the most common species in the examined towns was *E. helleborine*, reported in 21 out of 42 cities (Rewicz et al. 2017).

Figure 1. *Epipactis helleborine* growing between the paving slabs in front of a house. Ontario, Canada (photo by A. Rewicz, 2020).

POLLINATION BIOLOGY

Literature data showed that the role of the *E. helleborine* pollinators is played mainly by wasps (Vespidae) belonging to the order Hymenoptera, including *Vespula vulgaris*, *V. germanica* and *Vespa sylvestris* (Knuth 1909; Judd 1979) (Table 1). Research of Rewicz et al. (2017) indicates the presence of the analyzed pollinating insects in two types of habitats (anthropogenic and natural). In both types of habitats, members of Diptera and Hymenoptera clearly predominated, grouping respectively 41% and 52% of all noted pollinators in anthropogenic sites as well as 59% and 37% in natural sites (Rewicz 2015, 2017). However, research of Rewicz et al. (2017) has not confirmed the dominance of Vespidae pollinators among the pollinators of *E. helleborine* (Table 2). In populations inhabiting anthropogenic habitats, *E. helleborine* were reported as being most frequently visited by: *Apis mellifera* (observed 27 times), *Meliscaeva cinctella* (23) and *Dolichovespula saxonica* (13).

Table 1. The main groups of *Epipactis helleborine* pollinators according to the literature data

Author	Year	Insect species
Müller	1873	*Vespula germanica, V. vulgaris*
Darwin	1877	*Vespa sylvestris*
Knuth	1909	*Vespa vulgaris*
Schremmer	1961	*Dolichovespula sylvestris, Vespula vulgaris, Paravespula rufa*
Judd	1970	*Vespula arenaria, V. consobria, V. vidua*
Wiefelspütz	1970	*Dolichovespula sylvestris, Vespula austriaca, V. vulgaris, V. germanica, Paravespula sylvestris*
Vöth	1982	*Dolichovespula media, D. saxonica, D. sylvestris, Paravespula rufa, Vespula germanica, V. vulgaris*
Petit	1986	*Vespula germanica*
Müller	1988	*Apis mellifera, Dolichovespula norvegica, D. saxonica, D. sylvestris, D. media, Lasioglossum calceatum, Myrmica laevinodis, Pravespula rufa, Vespula vulgaris*
Jakubska et al.	2005	*Coccinella septempunctata, Vespula germanica, V. vulgaris, Apis mellifera, Psithyrus bohemicus, Bombus hortorum, B. hypnorum*
Veenendaal	2010	*Dolichovespula saxonica, D. media*
Rewicz et. al.	2015, 2017	*Culicidae, Episyrphus balteatus, Meliscaeva cinctella, Dolichovespula saxonica, Apis mellifera, Bombus* sp., *Dolichovespula media, Vespula vulgaris, Formicidae* sp.

Table 2. Identified insects in the analyzed natural and anthropogenic populations of *Epipactis helleborine* (Rewicz 2015)

Insect species	Anthropogenic populations	Natural populations
Pollinating insects		
Diptera		
Culicidae	**2**	**3**
Episyrphus balteatus	**10**	**2**
Meliscaeva cinctella	**23**	**86**
Hymenoptera		
Apis mellifera	**27**	
Bombus sp.		**8**
Dolichovespula media	**2**	
Dolichovespula saxonica	**13**	**27**
Vespula vulgaris	**9**	
Formicidae	**4**	
Optional pollinating insects		
Diptera		
Melanostoma mellinum		**2**
Platycheirus sp.	**2**	
Sphaerophoria scripta	**5**	
Hymenoptera		
Apidae	**3**	**5**
Eumenes coarctatus	**1**	
Polistes nimphus	**4**	
Visiting insects		
Coleoptera		
Harmonia axyridis	**1**	
Cerambycidae		**1**
Dasytes niger		**3**
Elateridae		**1**
Diptera		
Aedes sp.	**6**	
Hemiptera		
Himacerus apterus	**1**	
Hymenoptera		
Eudorylus sp.	**1**	
Ichneumonidae	**3**	**1**
Orthoptera		
Chorthippus apricarius	**1**	

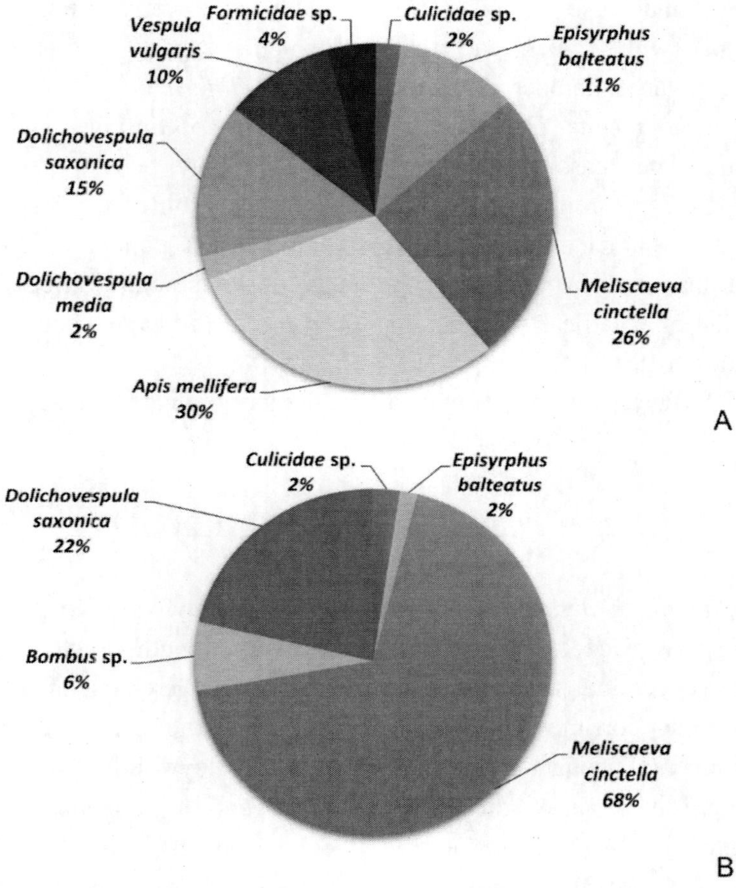

Figure 2. Percentage of visits by pollinating insects in the population in A) anthropogenic, B) natural habitats of *Epipactis helleborine* populations.

The populations from natural habitats were reported to be most frequently pollinated by: *Meliscaeva cinctella* (86 times), *Dolichovespula saxonica* (27) and *Bombus* sp. (8) (Figure 2) (Rewicz 2015; Rewicz et al. 2017). *M. cinctella*, considered to be the so-called universal pollinator of flowering plants, was found not only on plants of the family Orchidaceae, but also, among others, on Apiaceae, Asteraceae and Liliaceae (Chandler 1998). The high frequency of visits of *Episyrphus balteatus* (Syrphidae) to this species probably results from its wide range of occurrence (Jansen

2001). It may also be related to the color of flowers preferred by *E. balteatus* (white, green, and pink), on which it eagerly feeds (De Buck 1990). A large number of visits by *Apis mellifera* in "anthropogenic" populations results from the close proximity of apiaries and home gardens. The reasons for these differences may be the geographic location of the studied populations, the composition of the local insect pool, seasonal changes in the size of the insect population, weather conditions, and in particular the ambient temperature affecting the emission of volatile attractants contained in the *Epipactis* nectar (Ehlers and Olesen 1997).

The Figures 2A and 2B indicate the number of visits by insects.

EPIPACTIS HELLEBORINE VARIABILITY

Epipactis helleborine is a species with highly variable phenotype features (Rewicz et al. 2017). The shoot length, the length of inflorescence, and the leaf size are the most variable characteristics, and therefore support the view that such features are the most susceptible to environmental changes (Heslop-Harrison 1953; Heywood 1974, Table 3). Literature data also indicate that the flowers of *E. helleborine* are characterized by high variability in color and shape of the perianth element (Figure 3). Rewicz (2015) observed that within one population there were shoots with flowers of different color and lip shape (Figure 4).

Phenotypic plasticity of the species is an expression of its potential abilities to colonize areas which differ in terms of many habitat features (Sultan 1995, 2000, 2001).

Epipactis helleborine produces a great number of small and light seeds. Seeds of *E. helleborine* show phenotypic variability in seed size within a species, population, individual, or even fruit (Waller 1982; Cavers and Steele 1984; Hendrix 1984; Wulff 1986; Šerá and Šerý 2004). The seeds are spindle-shaped to elliptical and fusiform in shape. The seeds are made up of the seed coat and the embryo (Molvray and Kores 1995; Arditti and Ghani 2000; Akçin et al. 2009). The seed coat (pale brown to golden in color), often shiny, is membranous, thin, striated, and usually loosely packed around the embryo.

Table 3. Literature data on metric features of shoots of *Epipactis helleborine*

Character	Dimensions	References
Stem	4.4-149 cm	Raciborski and Szafer (1919); Hegi (1925); Weijer (1952); *Procházka* and *Velísek* (1983); Tyteca (1995); Delforge (2006); Bauman et al. (2010); Harrap and Harrap (2010); Bîtea et al. (2011); Rewicz (2015)
Number of leaves	4–15	Tutin et al. (1980); Delforge (2006)
Leaf length	1.7-17.5 cm	Tutin et al. (1980); Delforge (2006); Bîtea et al. (2011), Rewicz (2015)
Leaf width	1.5-13 cm	Tutin et al. (1980); Delforge (2006); Bîtea et al. (2011); Rewicz (2015)
Inflorescence length	2-57 cm	Delforge (2006); Bauman et al. (2010), Bîtea et al. (2011); Pipper and Waite (1988); Rewicz (2015)
The number of flowers	3-100	Delforge (2006)
Perianth leaves	8-13.5 mm long, 4.5-6 mm wide	Stepankova (2011); Klein (1997); Hegi (1925)

The seed coat is composed of elongated cells, 1.1 to 1.3 mm long and 0.2 mm wide (Swamy et al. 2004; Bojnansky and Fargasova, 2007; Akçin et al. 2009). Generally, the medial cells are longer than the apical and basal cells. The apical pole mainly consists of polygonal cells. The number of cells at the longest axis of the testa varies from 18 to 33 cells (Rewicz 2015).

The seeds vary in length from 0.15 mm to 1.3 mm and are from 0.20 mm to 0.76 wide (Table 4). The embryo consists of several dozen undifferentiated cells, most often spherical in shape. Its length is from 0.15 to 0.22 mm, and its width ranges from 0.11 to 0.19 mm (Table 4). The number of seeds in one boll varies from 1000 to 3000 (Arditti and Ghani 2000, Rewicz et al. 2016). Rewicz et al. (2017) reported that *E. helleborine* growing in natural habitats produced seeds heavier than in anthropogenic habitats (214 × 10-6 μg vs. 163 × 10-6 μg). However, the number of seeds per bag in anthropogenic habitats is greater than the number of seeds in natural habitats (2118 vs. 1813). The size of seeds, their weight and number are the result of adaptation to environmental conditions, and also depend on the life strategy of the plant (Salisbury 1942, 1975; Thompson 1987; Jankowska-Błaszczuk 1996; Bazzaz et al. 2000; Leishman et al. 2000; Faast et al. 2011).

Figure 3. Variability of perianth color and shape of *Epipactis helleborine* (photo by A. Rewicz, 2011 – 2013).

Variations in the number of seeds produced may result from *E. helleborine* adapting to various habitat conditions (Leishman et al. 2000), or as Light and MacConaill (1991) show in their research, they may be the result of climatic conditions that significantly reduce the number of seeds produced and may also result from unstable water conditions and high insolation (Salisbury 1975; Gibson and Gibson 2006). It is probably largely related to the life strategy of the species and environmental factors (Thompson 1987; Thompson et al. 1993; Jankowska-Błaszczuk 1996).

Figure 4. Intra-population variability of color, size and shape of elements of the perianth of *Epipactis helleborine* (photo by A. Rewicz, 2013).

Fruits can be formed through allogamy or autogamy (Tałałaj and Brzosko 2008). Rewicz et al. (2017) showed that the mean number of fruits produced in autogamy in the "anthropogenic" population was significantly higher than the number of fruits in the natural population.

Table 4. Literature data on *Epipactis helleborine* seed metrics

Character		Study					
		Young 1962	Arditti et al. 1980	Richards and Porter 1982	Arditti and Ghani 2000	Bojnansky and Fargasova 2007	Rewicz et al. 2016
Seed	length [mm]	0.97	0.99-1.29	0.85-0.91	1.15-0.15	1.10-1.30	0.53-0.82.
	width [mm]	0.27	0.21-0.22	-	0.25-0.02	0.2	0.56-0.76
	weight [µg]	-	-	-	-	-	$163 \times 10^{-6-}$ 214×10^{-6}
Embryo	length [mm]	-	0.19-0.22	0.15	-	-	0.17-0.21
	width [mm]	-	0.11-0.12	0.15	-	-	0.18-0.19

LIFE CYCLE AND SPATIAL STRUCTURE

A response of a plant to environmental conditions can manifest itself in its morphological variability, physiological responses or reproductive potential. *Epipactis helleborine* can reproduce generatively or vegetatively. Both strategies are usually implemented in populations, however, depending on individual habitat differences, their mutual proportions may be different (Procházka and Velisek 1983; Stefaniak et al. 2011).

Rewicz (2015) showed that the distribution of *E. helleborine* shoots in anthropogenic populations was cluster-like, with the largest groupings of shoots observed there (6 to 12 shoots/1 m^2) (Figure 5). Most of the *E. helleborine* shoots in ″anthropogenic″ populations reappeared in the same place in the next research season. New shoots (especially vegetative ones) most often appeared near the largest (oldest) generative shoots (Figure 6). The observed distribution of shoots in the studied anthropogenic populations may indicate the dominance of vegetative multiplication – a clonal type of growth (Rewicz et al. 2018). In ″anthropogenic″ populations, the possibility of carrying out vegetative multiplication may affect the appearance of shoots in the same place in subsequent vegetation seasons, the appearance of new shoots in close proximity to the previous ones, and an increase in shoot density in a given area. This is confirmed, among others, by research conducted by Wyrzykiewicz-Raszewska (2001) and Light and MacConnail (2006). Falińska (2002) in her research on the marsh thistle showed that in the case of clonal plants, individuals may multiply through rhizome fragmentation, resulting in new shoots being produced over time. Fragmentation of shoots may result, among others, from defense against infection or pathogens (Kelly 1995). In turn, Grime (1979) points to the importance of genetic disintegration preventing plants from aging through the use of vegetative expansion. In the case of species of the genus *Epipactis*, it is also supported by their perennial strategy, which enables the longevity of the gene by replacing old structural units with new ones (Harper and Ogden 1970).

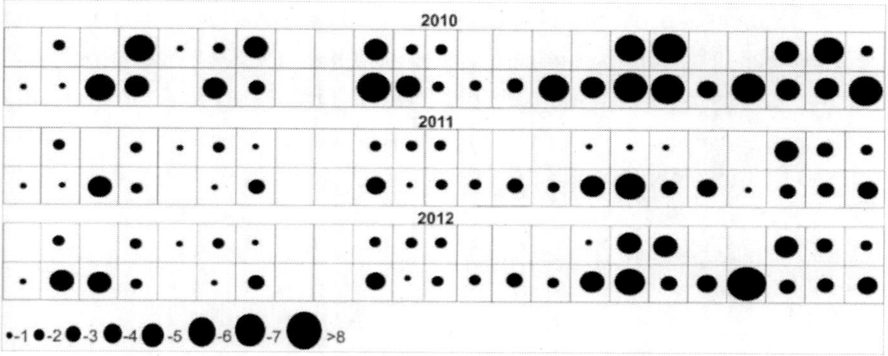

Figure 5. Changes in the density of *Epipactis helleborine* shoots in the anthropogenic population in Guszczewina (Poland). The size of the dots means the number of shoots in a square of 1×1 m.

Figure 6. *Epipactis helleborine* shoots growing side by side (photo by A. Rewicz, Białowieża Primeval Forest, 2012).

In the case of natural populations, the shoots did not form as distinct clusters, as it was in the anthropogenic populations. The populations were characterized by a low density per 1 m^2 (2 to 3 shoots /1 m^2) and the emerging new shoots were not observed in close proximity to the tallest shoots or those observed in the previous season (Figure 7).

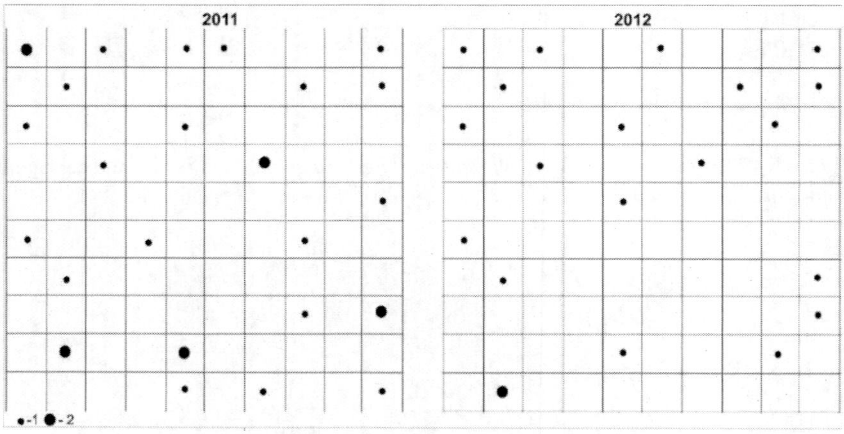

Figure 7. Changes in the density of *Epipactis helleborine* shoots in the natural population in the Kaczawskie Mountains (Poland). The size of the dots means the number of shoots in a square of 1×1 m.

Rewicz (2015) showed that flowers in natural populations produced a greater amount of live pollen (87%) compared to flowers from anthropogenic populations (67%). The research has shown that the mean number of fruit sets by autogamy in the "anthropogenic" population (20) was significantly higher than the number of fruit sets in the natural population (12).

Fungal Associations

As in all orchids, germination of *E. helleborine* seeds depends on the symbiosis with mycorrhizal fungus (Arditti and Ghani 2000). This species is partially myco-heterotrophic also in the adult stage (Bidartondo et al. 2004). Previous studies conducted have showed that *E. helleborine* has a broad spectrum of mycorrhizal partners including *Ceratobasidium*, *Tuber*, *Wilcoxina*, and several sebacinoids (Bidartondo et al. 2004; Ogura-Tsujita and Yukawa 2008; Těšitelová et al. 2012; Jacquemyn et al. 2016; Ogórek et al. 2020). The diversity of mycorrhizal partners in the case of the broad-leaved helleborine is one of the factors that affect the successful colonization of various habitats.

Genetic Variation

Squirell et al. (2001) analyzed the difference in genetic variation between European and Canadian populations of *E. helleborine* and found no evidence of a genetic bottleneck associated with the introduction of *E. helleborine* to North America. Allozymes and cpDNA RFLPs showed equivalent or higher levels of intrapopulation genetic diversity and lower levels of interpopulation differentiation in introduced populations relative to native populations (Squirell et al. 2001). The genetic variation between urban and rural *E. helleborine* populations was investigated by Hollingsworth and Dickson (1997). They showed in their research that the "large urban populations actually have patterns of variation similar to rural populations and show evidence of multiple founders."

Conclusion

Aims to review and analyze the characteristics of *Epipactis helleborine* (L.) Crantz, one of the most widespread orchids of the Temperate Zone that allow it to grow in different kinds of habitats. The authors point out that the potential of *E. helleborine* to occupy new habitats and spread is due to its enormous phenotypic plasticity, as well as to a wide range of mycorrhizal symbionts and pollinating insects (Figure 7). Thanks to its remarkably high adaptability to various environmental conditions, this taxon can be classified as an "undemanding species". In fact, such a configuration of characteristics makes it a "super species" that will surprise us with the conquest of new habitats and territories in the future.

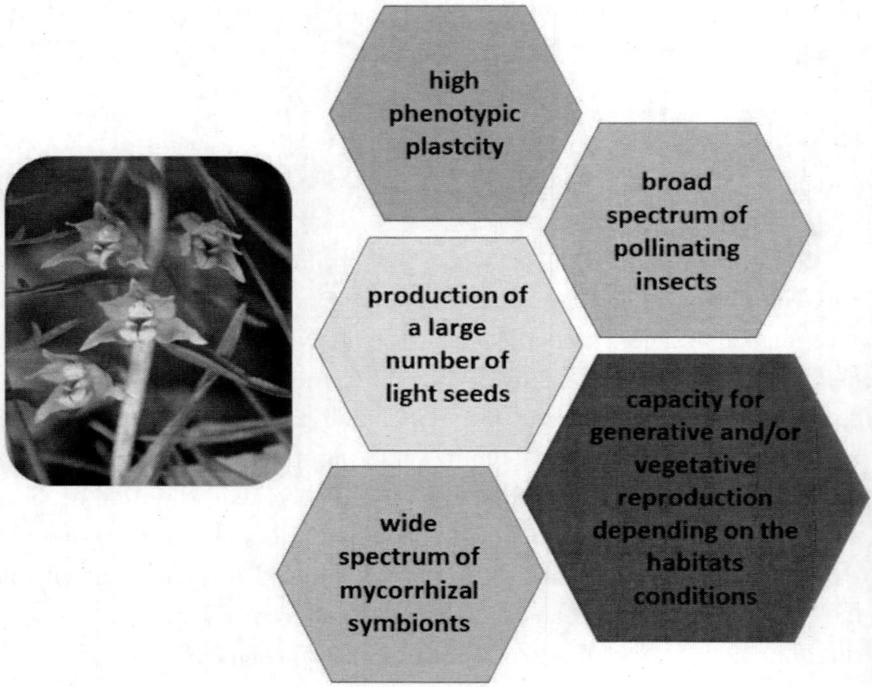

Figure 8. The *Epipactis helleborine* "attributes" allow for its widespread.

Most plants which are classified as pioneers and potentially invasive are characterized by rapid growth and maturity, abundant seed production, highly successful seed dispersal, germination and colonization, as well as the possibility of vegetative spread (Tiwari et al. 2005). *Epipactis helleborine* has almost all these attributes apart from the first one. The development from a seed to a seedling with an aerial shoot may take 3–5 years (Vakhrameeva et al. 2008; Hemrová et al. 2019), however, the seed bank is rather long-lived (Whigham et al. 2006).

Numerous infraspecific taxa have been described within *E. helleborine* based on morphological characters (e.g., Kreutz 2005, 2007; Riechelmann and Zirnsack 2008), however, due to the enormous phenotypic plasticity of this species, their actual separateness is difficult to evaluate. A broad-scale genetic study could help to evaluate the actual gene flow between populations as well as the status of proposed subspecies and varieties. The proper assessment of this taxonomic diversity could help to identify forms of *E. helleborine* which are more widespread and have larger invasive potential.

References

Adamowski, Wojciech. "Expansion of Native Orchids in Anthropogenous Habitats." *Polish Botanical Studies* 22 (2006): 35–44.

Adamowski, Wojciech. "Phenotypic Variation of *Epipactis helleborine* × *E. atrorubens* Hybrids in Anthropogenic Conditions." *Acta Societatis Botanicorum Poloniae* 64, no. 3 (1995): 303–12.

Adamowski, Wojciech. *Apofityzm wybranych gatunków storczykowatych (*Orchidaceae*) i jego uwarunkowania ekologiczne,* [*Apophytism of Chosen Species of* Orchidaceae *and Its Ecological Conditions*]. PhD Thesis, Warsaw University, Warszawa, 1996.

Adamowski, Wojciech. "Colonization Success of Orchids in Disturbed Habitats". In *Plant Population Biology and Vegetation Processes* edited by Falińska K. Kraków: Institute of Botany PAS, 1998.

Akçin, Tülay Aytaç, Yasemin Ozdener, and Adnan Akçin. "Taxonomic Value of Seed Characters in Orchids from Turkey." *Belgian Journal of Botany* 142, no. 2 (2009): 124–39.

Akhalkatsi, Maia, Giorgi Arabuli, and Richard Lorenz. "Orchids as Indicator Species of Forest Disturbances on Limestone Quarry in Georgia (South Caucasus)". *Journal Europäischer Orchideen* 46, no. 1 (2014): 123–60.

Arditti, Joseph, and Abdul Karim Abdul Ghani. "Numerical and Physical Properties of Orchid Seeds and Their Biological Implications". *New Phytologist* 145 (2000): 367–421.

Arditti, Joseph, P. Ernst, T. W. Yam, and C. Glabe. "The Contributions of Orchid Mycorrhizal Fungi to Seed Germination: a Speculative Review. *Lindleyana* 5, no. 4 (1990): 249–55.

Arditti, J., J. D. Michaud, and P. L. Healey. "Morphometry of Orchid Seeds. Ii. Native California and Related Species of *Calypso*, *Cephalanthera*, *Corallorhiza* and *Epipactis*."*American Journal of Botany* 67, no. 3 (1980): 347–60.

Bazzaz, A. F., D. D. Ackerly, and G. E. Reekie. "Reproductive Allocation in Plants". In *CAB International. Seeds: The Ecology of Regeneration in Plant Communities*, 12nd edition, edited by M. Fenner. UK: School of Biological Sciences University of Southampton, 2000.

Baumann, H., S. Kunkele, and R. Lorenz. *Storczyki Europy i obszarów sąsiednich* [*Orchids of Europe and neighboring areas*]. Warszawa: Multico, 2010.

Berdowski, W., and K. Spałek. „Rozmieszczenie oraz zasoby chronionych gatunków roślin naczyniowych we wschodniej części województwa opolskiego [Distribution of Resources and Protected Species of Vascular Plants in the Eastern Part of the Province of Opole]". *Acta Universitatis Wratislaviensis*1936, *Prace Botaniczne* 73 (1997): 7–26.

Berezutsky, M. A., T. B. Reshetnikova L. A. Serova, and A. S. Kashin. "Ecological Despecialization of Species of the Family Orchidaceae Juss. in the North of the Lower-Volga Region". *Biology Bulletin* 41, no. 10 (2014): 849–50.

Bernacki, Leszek, and Błońska Agnieszka. "Występowanie storczykowatych (Orchidaceae) w otoczeniu południowego odcinka drogi Krośnica–Niedzica w Pienińskim Parku Narodowym, [Occurrence of Orchids (Orchidaceae) in the Vicinity of the Southern Section of Krośnica–Niedzica Road in the Pieniny National Park]". *Pieniny – Przyroda i Człowiek* 9 (2006): 65–70.

Bidartondo, Martin I., Bastian Burghardt, Gerhard Gebauer, Thomas D. Bruns, and David J. Read. "Changing Partners in the Dark: Isotopic and Molecular Evidence of Ectomycorrhizal Liaisons between Forest Orchids and Trees." *Proceedings. Biological Sciences* 271, no. 1550 (2004): 1799–1806.

Bîtea, Nicoleta Daniela, E. Madoşa, S. Ciulcă, S. Popescu, A. Lazăr, C. Avădanei, and G. Velicevici. "Studies Regarding the Morphological Variability of Some Orchid Species from Anina Mountains." *Journal of Horticulture Science and Forestry* 5, no. 2 (2011): 43–47.

Bojnansky, V., and A. Fargasowa. *Atlas of Seeds and Fruits of Central and East- Europen Flora*. Springer, 2007.

Brown, P. M. and G. W. Argus. "*Epipactis* Zinn, Orchidaceae". *North American flora* 26 (2002): 584–586.

Cavers, Paul B., and Marion G. Steel. "Patterns of Change in Seed Weight over Time on Individual Plants." *The American Naturalist* 124, no. 3 (1984): 324–35.

Chandler. P. J. *Checklists of Insects of the British Isles (New Series). Part 1: Diptera*. Handbooks for the Identification of British Insects, 1998.

Cozzolino, Salvatore and Alex Widmer. "Orchid Diversity: an Evolutionary Consequence of Deception?". *Trends in Ecology and Evolution* 20, no. 9 (2005): 487–94.

Ćwikliński, E. "Flora synantropijna Zielonej Góry i Koszalina na tle warunków przyrodniczych i rozwoju miast [Synanthropic Flora of the Zielonej Mountains and Koszalin Against Natural Conditions and Urban Development]". *Materiały Zakładu Fitosociologii Stosowanej UW* 27 (1971): 81–113.

Darwin, C. *The Effects of Cross and Self Fertilization in the Vegetable Kingdom.* London, 1877.

Day, D. "*Epipactis helleborine* in Erie County". *Bulletin of the Torrey Botanical Club* 9 (1882): 127–28.

De Buck, N. "Bloembezoek en bestuivingsecologie van Zweefvliegen (Diptera, Syrphidae) in het bijzonder voor België [Flower Visits and Pollination Ecology of Hoverflies (Diptera, Syrphidae) in Particular for Belgium]": *Documents de travail de l'Institut Royal des Sciences naturelles de Belgique,* 60:3-167. Brussel: Studiedocumenten K.B.I.N., 1990.

Delforge, Pierre. *Guide des Orchidées d'Europe, d'Afrique du Nord et du Proche-Orient* [*Orchids of Europe, North Africa and the Middle East*]: 3e éd. Paris: Delachaux et Niestlé, 2006.

Dickson, J. H. "*Epipactis helleborine* in Gardens and Other Urban Habitats: an Example for Apophytism." in *Urban Ecology*, edited by H. Sukopp and S. Hejny. The Hague: SPB Academic Publishing, 1990.

Djordjević, Vladan, Spyros Tsiftsis, Dmitar Lakušić, Slobodan Jovanović, Ksenija Jakovljević, and Vladimir Stevanović. "Patterns of Distribution, Abundance and Composition of Forest Terrestrial Orchids." *Biodiversity and Conservation* 29, no. 14 (2020a): 4111–34.

Djordjević, Vladan, Spyros Tsiftsis, Dmitar Lakušić, Slobodan Jovanović, and Vladimir Stevanović. "Orchid Species Richness and Composition in Relation to Vegetation Types." *Wulfenia* 27 (2020b): 183–210.

Ehlers, Bodil K., and Jens M. Olesen. "The Fruit-Wasp Route to Toxic Nectar in *Epipactis* Orchids?" *Flora* 192, no. 3 (1997): 223–29.

Evans, Alexandra, and Hans Jacquemyn. "Impact of Mating System on Range Size and Niche Breadth in *Epipactis* (Orchidaceae)." *Annals of Botany* 126, no. 7 (2020): 1203–14.

Faast, R., Facelli J. M. and A. D. Austin. "Seed Viability in Declining Populations of *Caladenia rigida* (Orchidaceae): Are Small Populations Doomed?" *Plant Biology* 13, no. 1 (2011): 86–95.

Falińska, Krystyna. *Ekologia roślin* [*Plant Ecology*]. Warszawa: Wydawnictwo Naukowe PWN, 2002.

Fay, Michael F. "Orchid Conservation: How Can We Meet the Challenges in the Twenty-First Century?" *Botanical Studies (Taipei, Taiwan)* 59, no. 1 (2018): 16.

Geisselbrecht-Taferner, L., Mucina L. "*Bidentetea tripartitae*". In *Die Pflanzengesellschaften Österreichs – Teil I*, edited by L. Mucina, L.

Geisselbrecht-Taferner, T. Ellmauer. Jena: Anthropogene Vegetation, 1993. [*The Plant Societies of Austria - Part I*]

Godefroid, S. "*Epipactis helleborine* en Extension à Bruxelles, [*Epipactis helleborine* extended in Brussels]". *Adoxa* 6/7 (1995): 13–14.

Gravendeel, Barbara, Ann Smithson, Ferry J. W. Slik, and Andre Schuiteman. "Epiphytism and Pollinator Specialization: Drivers for Orchid Diversity?". *Philosophical Transactions of the Royal Society of London. Series B, Biological Sciences* 359 (2004): 1523–35.

Gibson, J. P., and T. R. Gibson. *Plant Ecology*. Philadelphia: Chelsea House Publishers, 2006.

Grime, John Philip. *Plant Strategies and Vegetation Processes*. Chichester, UK: Wiley, 1979.

Harper, John L., and John Ogden. "The Reproductive Strategy of Higher Plants. I. The Concept of Strategy with Special Reference to *Senecio vulgaris* L". *Journal of Ecology* 58, no. 3 (1970): 681–98.

Harrap, Anne and Simon Harrap. *Orchids of Britain and Ireland*. London: A & C BLACK, 2010.

Hegi, Gustav. *Illustrierte Flora von Mittel-Europa*. 2. München: J. F. Lehmann Verlag, 1925. [*Illustrated flora of Central Europe.*]

Hemrová, Lucie, Milan Kotilínek, Marie Konečná, Radim Paulič, Jana Jersáková, Tamara Těšitelová, Jana Knappová, and Zuzana Münzbergová. "Identification of Drivers of Landscape Distribution of Forest Orchids Using Germination Experiment and Species Distribution Models." *Oecologia* 190, no. 2 (2019): 411–23.

Hendrix, Stephen D. "Variation in Seed Weight and Its Effects on Germination in *Pastinaca sativa* L. (Umbelliferae)". *American Journal of Botany* 71 (1984): 795–802.

Heslop-Harrison, John. *New Concepts in Flowering-plant Taxonomy*. London: Heinemann, 1953.

Heywood, H. Vernon. *Taksonomia roślin*. Warszawa: Wydawnictwo Naukowe PWN, 1974.

Hollingsworth, P. M., and J. H. Dickson. "Genetic Variation in Ruderal and Urban Populations of *Epipactis helleborine* (L.) Crantz

(Orchidaceae) in Britain". *Biological Journal of the Linnean Society* 123 (1977): 321–31.

Hultén E., and M. Fries. *Atlas of North European Plants North of the Tropic of Cancer*. Königstein: Koeltz Scientific Books, 1986.

Jackowiak, Bogdan. *Dynamics Synanthropisation Flora of Poznan as a Result of Anthropogenic Changes in the Ecosystem of a Big City*. Part 2. Wykaz i charakterystyka roślin naczyniowych Poznania, Poznań: Adam Mickiewicz University, 1986.

Jacquemyn, Hans, Michael Waud, Bart Lievens, and Rein Brys. "Differences in Mycorrhizal Communities between *Epipactis palustris, E. helleborine* and Its Presumed Sister Species *E. neerlandica". Annals of Botany* 118, no. 1 (2016): 105–14.

Jakubska, Anna, Maria Malicka, and Marek Malicki. "New Data on the Apophytic Occurrence of *Epipactis helleborine* (L.) Crantz and *Cephalanthera longifolia* (L.) Fritsch in *Populus × canadensis* Plantation in Lower Silesia (South-Western Poland)". *Biodiversity: Research and Conservation* 1, no. 2 (2006): 96–98.

Jakubska, Anna, Marcin Kadej, Dorota Prządo, and Mieczysław Steininger. "Ekologia zapylania *Epipactis helleborine* (L.) Crantz (Orchidaceae, Neottieae) w południowo-zachodniej Polsce". *Acta Botanica Silesiaca* 2 (2005): 131–44. [Ecology of pollination *Epipactis helleborine* (L.) Crantz (Orchidaceae Orchidaceae, Neottieae) in south-west Poland]

Jankowska-Błaszczuk, Małgorzata. "Ekologiczne znaczenie wielkości nasion". *Wiadomości Botaniczne* 40 (1996): 19–30. [Ecological importance of seed size. *Botanical News*]

Judd, W. W. "Arthropods associated with Helleborine Orchid, *Epipactis helleborine*". *Entomology News* 90, no.1 (1979): 41–44.

Jurkiewicz, A., K. Turnau, J. Mesjasz-Przybylowicz, W. Przybylowicz, and B. Godzik. "Heavy Metal Localisation in Mycorrhizas of *Epipactis atrorubens* (Hoffm.) Besser (Orchidaceae) from Zink Mine Tailings". *Protoplasma* 218 (2001): 117–24.

Kelly, C. K. "Thoughts on Clonal Integration – Facing the Evolutionary Context". *Ecology and Evolution* 9, no. 6 (1995): 575–85.

Kiedrzyński, Marcin, and Agnieszka Stefaniak. "Występowanie rodzaju *Epipactis* Zinn 1757 w Puszczy Pilickiej na stanowiskach naturalnych i antropogenicznych". In *Storczykowate w Polsce Biologia i Ochrona*, edited by E. Brzosko, A. Wróblewska, and E. Jarmakowicz, 86–96. Białystok, 2011. [The occurrence of the genus Epipactis Zinn 1757 in the Pilica Forest at natural and anthropogenic positions. In *Orchids in Poland Biology and Protection*]

Klein, Erich. "*Epipactis helleborine* (L.) Crantz subsp. *orbicularis* (Richter) Klein comb. nova, eine xerophile Unterart (Orchidaceae-Neottieae)". *Phyton* 37 (1997): 71–83. [*Epipactis helleborine* (L.) Crantz subsp. *orbicularis* (Richter) Small comb. nova, a xerophilic subspecies (Orchidaceae-Neottieae)]

Klimko, Małgorzata, Aneta Czarna, and Bogdan Bałuka. "Vascular Plants of Post-industrial Habitats in Wałbrzych City". *Acta Botanica Silesiaca* 1 (2004): 7–22.

Knuth, *Paul. Handbook of Flower Pollination: Based Upon Hermann Müller's Work 'The Fertilisation of Flowers by Insects'*. Oxford: Clarendon Press, 1909.

Kolanowska, Marta. "Niche Conservatism and the Future Potential Range of *Epipactis helleborine* (Orchidaceae)". *PLoS ONE* 8, no. 10 (2013): e77352.

Kreutz, Carolus Adrianus Johannes. "Korrekturen und Ergaenzungen zum Kompendium der Europaeischen Orchideen". *Eurorchis* 17 (2005): 97–128. [Corrections and additions to the Compendium of European Orchids]

Kreutz, Carolus Adrianus Johannes. "Beitrag zur Kenntnis europäischer, mediterraner und vorderasiatischer Orchideen". *Berichte aus den Arbeitskreisen Heimische Orchideen* 27, no. 2 (2010): 171–236. [Contribution to the knowledge of European, Mediterranean and Near Eastern orchids. *Reports from the local orchids working groups*]

Leishman, M. R., I. J. Wright, A. T. Moles, and M. Westoby. "The Evolutionary Ecology of Seed Size." In *Seeds – the Ecology of*

Regeneration in Plant Communities, edited by M. Fenner, 2nd edition. Wallingford: CAB International, 2000.

Light, M. H. S., and Michael MacConaill. "Patterns of Appearance in *Epipactis helleborine* (L.) Crantz." in *Population Ecology of Terrestrial Orchids*, edited by T. C. E. Wells, and J. H. Willems. The Netherlands: The Hague Academic, 1991.

Light, Marilyn H. S., and Michael MacConaill. "Appearance and Disappearance of a Weedy Orchid, *Epipactis helleborine*." *Folia Geobotanica* 41, no. 1 (2006): 77–93. McCormick, Meliisa K., and Hans Jacquemyn. "What Constrains the Distribution of Orchid Populations?" *New Phytologist* 202 (2014): 392–400.

Milović, Milenko, and Božena Mitić. "Urbana Flora Zadra (Dalmacija, Hrvatska) [Urban Flora of Zadar (Dalmacija, Hrvatska]". *Natura Croatica* 21 (2012): 65–100.

Misiewicz, Jan. *Flora synantropijna Słupska na tle warunków przyrodniczych i rozwoju miasta* [*Synanthropic Floraof Slupsk on the Background of Natural Conditions and Development of the City*]. Słupsk: WSP, 1978.

Molvray, M., and J. P. Kores. "Character Analysis of the Seed Coat in Spiranthoideae with Special Reference to the Diurideae (Orchidaceae)". *American Journal of Botany* 82 (1995): 1443–53.

Müller, Hermann. *Die Befruchtung der Blumen durch Insekten und die gegenseitige Anpassung beider* [*The Fertilization of Flowers by Insects and Their Mutual Adaptation for that Function*]. Cambridge: Engelmann Verlag, Leipzig, 1873.

Müller, I. "Vergleichende blutenokologische Untersuchungen an der Orchideengattung *Epipactis". Mitt Bl Arb Kr Heim Orchid Baden-Wurtt,* 20 (1988): 701–803. ["Comparative blood ecology studies on the orchid genus Epipactis". *Mitt Bl Arb Kr Heim Orchid Baden-Wurtt*]

Nikolaeva, V. M., and B. M. Zefirov. *Flora Belovezhskoi pushchi (sosudistye sporovye i semennye rasteniya).* Minsk: Izdatelstvo Uradzhai, 1971. [*Flora of Belovezhskaya Pushcha (vascular spore and seed plants).*]

Nowak, Teresa. "Flora synantropijna linii kolejowej Dąbrowa Górnicza – Strzemieszyce – Olkusz. [Synanthropic flora of railway in Dąbrowa Górnicza – Strzemieszyce – Olkusz]". *Acta Biologica Silesiaca* 30, no. 47 (1997): 86–105.

Ogura-Tsujita, Y., and T. Yukawa. "*Epipactis helleborine* Shows Strong Preference Towards Ectomycorrhizal Fungi with Contrasting Geographic Distributions in Japan". *Mycorrhiza* 18 (2008): 331–38.

Ogórek, Rafał, Klaudia Kurczaba, Zbigniew Łobas, Elżbieta Żołubak, and Anna Jakubska-Busse. "Species Diversity of Micromycetes Associated with *Epipactis helleborine* and *Epipactis purpurata* (Orchidaceae, Neottieae) in Southwestern Poland". *Diversity* 12, no. 5 (2020): 182.

Owen, Maria L. "An Orchid New to America". *The Bulletin of the Torrey Botanical Club* (1879): 329–30.

Petit, J. "Quelques observations sur les insectes butineurs d' *Epipactis helleborine* (L.) Crantz [Some observations on the foraging insects of *Epipactis helleborine* (L.) Crantz]". *Dumortiera* 34–35 (1986): 112–16.

Piotrowska, Halina. "Rośliny naczyniowe wysp Wolina i południowo-wschodniego Uznamu. [Vascular Plants of the islands Wolin and south-eastern Usedom]". *Prace Komisji Biologicznej PoznańskiegoTowarzystwa Przyjaciół Nauki* 30, no. 4 (1966): 1–282. [*Works of the Biological Commission of the Poznań Society of Friends of Science*]

Piper, J., and S. Waite. "The Gender Role of Flowers of Broad Leaved Helleborine *Epipactis helleborine* (L.)". *Functional Ecology* 2 (1998): 35–40.

Procházka, F., and Velísek V. *Orchideje nasi prirody* [*Orchids of Our Nature*], Prague: Ceskoslovenska Akademie, 1983.

Raciborski, Marjan and Władysław Szafer. *Flora polska roślin naczyniowych i ziem ościennych. tom 1. Paprotniki iglaste i jednoliścienne* [*Polish flora of vascular plants and neighboring lands. vol 1. Coniferous and monocotyledonous ferns*]. Kraków: Akademja Umiejętności, 1919.

Rebele, F. "Ergebnisse floristischer Untersuchungen in den Industriegebieten von Berlin (West), [Results of Floristic Studies in the Industrial Areas of Berlin]". *Landschaft und Stadt* 20, no. 2 (1988): 49–66. [*Landscape and city*]

Remmel, G. "Ungewöhnliche *Epipactis*-Formen auf der Halde eines alten Erzbergwerks [*Epipactis* in the old iron ore mine]". *Journal ber Naturwiss Ver Wuppertal* 23 (1970): 119–22.

Renz, J., S. Ali, and E. Nasir. *Flora of Pakistan No. 164*, Orchidaceae. Karachi, Pakistan: University of Karachi, 1984.

Rewicz, Agnieszka, Anna Bomanowska, Myroslav V. Shevera, Józef K Kurowski, Katarzyna Krasoń, and Katarzyna M. Zielińska. "Cities and Disturbed Areas as Man-made Shelters for Orchid Communities". *Notulae Botanicae Horti Agrobotanici Cluj-Napoca* 45, no. 1 (2017): 126–39.

Rewicz, Agnieszka, Radomir Jaskuła, Tomasz Rewicz, and Grzegorz Tończyk. "Pollination Biology and Reproductive Success of *Epipactis helleborine* (L.) Crantz (Orchidaceae) in Anthropogenic and Natural Habitats". *Peer J* 5 (2017): 1–24.

Rewicz, Agnieszka, Jeremi Kołodziejek, and Anna Jakubska-Busse. "The Role of Anthropogenic Habitats as Substitutes for Natural Habitats: a Case Study on *Epipactis helleborine* (L.) Crantz (Orchidaceae, Neottieae). Variations in Size and Nutrient Composition of Seed". *Turkish Journal of Botany* 40, no. 3 (2016): 258–68.

Rewicz, Agnieszka, Monika Rewers, Iwona Jędrzejczyk, Tomasz Rewicz, Jeremi Kołodziejek, and Anna Jakubska-Busse. "Morphology and Genome Size of *Epipactis helleborine* (L.) Crantz (Orchidaceae) Growing in Anthropogenic and Natural Habitats". *PeerJ* 2018: 6:e5992.

Rewicz, Agnieszka. *Biologia i ekologia Epipactis helleborine (L.) Crantz (*Orchidaceae*) na siedliskach naturalnych i antropogenicznych.* PhD Dissertation, Poland: Univeristy of Lodz, 2015. [*Biology and ecology of Epipactis helleborine (L.) Crantz (*Orchidaceae*) in natural and anthropogenic habitats"*]

Richards, A. J., and A. F. Porter. "On the Identity of a Northumberland *Epipactis*". *Watsonia* 14 (1982): 121–28.

Riechelmann, A., and A. Zirnsack. "*Epipactis helleborine* (L.) Crantz subsp. *moratoria* A. Riechelmann & A. Zirnsack, eine neue *Epipactis*-Unterart aus der nördlichen Fränkischen Alb". *Berichte aus den Arbeitskreisen Heimische Orchideen* 25, no. 1 (2008): 57–84. ["*Epipactis helleborine* (L.) Crantz subsp. *moratoria* A. Riechelmann & A. Zirnsack, a new *Epipactis* subspecies from the northern Franconian Alb". *Reports from the local orchids working groups*]

Salisbury, E. J. *The Reproductive Capacity of Plants*. Bell, London, 1942.

Salisbury, E. J. 1975. "Seed Size and Mass in Relation to Environment". *Proceedings of the Royal Society, London B* 186 (1975): 83–88.

Sauerland, K. E. "Orchideen in Rostocker Grünanlagen. Naturschutzarb". *Mecklenburg-Vorpommern* 35, no. 2 (1995): 53–54.

Schremmer, F. "Bemerkenswerte Wechselbeziehungen zwischen Orchideeblüten und Insekten". *Naturu Volk* 91 (1961): 52–61. [Remarkable Interrelationships Between Orchid Flowers and Insects]

Šerá, Božena, and Michal Šerý. "Relation Between Number and Weight of Seeds and Reproductive Strategies in Herbaceous Plants". *Folia Geobotanica* 39 (2004): 27–40.

Sinker, C. A., J. R. Packham, I. C. Trueman, P. H. Oswald, F. H. Perring, and W. Prestwood. *Ecological Flora of the Shropshire Region*. Shrewsbury: Shropshire Wildlife Trust, 1991.

Soczewka, B. "New Localities of Rare Vascular Plant Species in the Mesoregion of Łomaska Depression (eastern Poland)". *Fragmenta Floristica et Geobotanica* 7 (2000): 81–91.

Squirrell, J., P. M. Hollingsworth, R. M. Bateman, J. H. Dickson, M. H. Light, M. Macconaill, and M. C. Tebbitt. "Partitioning and Diversity of Nuclear and Organelle Markers in Native and Introduced Populations of *Epipactis helleborine* (Orchidaceae)." *American Journal of Botany* 88, no. 8 (2001): 1409–18.

Stepankova, J. *Flora Czech Kvetena*, tom 8. Praha: Academia, 2011.

Stešević, Danijela, and S. Jovanović. "Flora of the City of Podgorica, Montenegro (Taxonomic Analysis)". *Archives of Biological Sciences* 60 (2008): 245–53.

Stolarz, P. "Kruszczyk szerokolistny *Epipactis latifolia* na torowisku kolejki leśnej w lasach. [*Epipactis latifolia* on Guideway Forest Railway in Okuniew Rembertów Forest]". *Chrońmy Przyrodę Ojczystą* 50, no. 5 (1994): 73–74.

Sudnik-Wójcikowska, Barbara. "Flora Warszawy i jej przemiany w ciągu XIX I XX wieku". [*Flora* of the City of *Warsaw and Its Transformation During the* XIX and XX Centuries]. 1-2. Warszawa: UW, 1987.

Suetsugu, Kenji. "Delayed Autonomous Self-pollination in Two Japanese Varieties of *Epipactis helleborine* (Orchidaceae)". *Botanical Journal of the Linnean Society* 173 (2013): 733–43.

Sultan, Sonia E. "Phenotypic Plasticity and Plant Adaptation". *Acta Botanica Neerlandica* 44 (1995): 363–83.

Sultan, Sonia E. 2000. "Phenotypic Plasticity for Plant Development, Function and Life History". *Trends in Plant Science* 5 (2000): 537–42.

Sultan, Sonia E. 2001. "Phenotypic Plasticity for Fitness Components in *Polygonum* Species of 544 Contrasting Ecological Breadth". *Ecology* 82 (2001): 328–43.

Süveges, Kristóf, Viktor Löki, Ádám Lovas-Kiss, Tibor Ljubka, Réka Fekete, Attila Takács, Orsolya Vincze, Balázs András Lukács, and Attila Molnár V. "From European Priority Species to Characteristic Apophyte: *Epipactis tallosii* (Orchidaceae)." *Willdenowia* 49, no. 3 (2019): 401–9.

Swamy, K., H. N. Krishna Kumar, T. M. Ramakrishna, and S. N. Ramaswamy. 2004. "Studies on Seed Morphometry of Epiphytic Orchids from Western Ghats of Karnataka". *Taiwania* 49, no. 2 (2004): 124–40.

Swarts, D. Nigel, and Kingsley W. Dixon. "Terrestrial Orchid Conservation in the Age of Extinction". *Annals of Botany* 104 (2009): 543–56.

Świercz, Anna. "The Cement and Lime Industry and the Occurrence of Selected Species of Orchids". *Przegląd Przyrodniczy* 15, no. 3–4 (2009): 117–23.

Świercz, Anna. "Storczyki a przemysł cementowy w regionie świętokrzyskim, [Orchids and Cement Industry in the Region Świętokrzyskie]". *Parki Narodowe* 2 (2005): 24–26.

Świercz, Anna. Przegląd stanowisk wybranych gatunków z rodziny Orchidaceae w sąsiedztwie cementowni regionu świętokrzyskiego, [*Survey of Selected Species from the Family* Orchidaceae *in the Vicinity of the Cement Plant in the Region of Kielce*]. Warszawa: Regionalne Studia Ekologiczno-Krajobrazowe, Problemy Ekologii Krajobrazu, 2006.

Świercz, Anna. "Wpływ cementowni Ożarów (Przedgórze Iłżeckie) na zmianę składu gatunkowego zbiorowisk borowych w latach 1993-2003 [Effect of Ożarów Cement Plant (*Przedgórze Iłżeckie*) on the Changes Floristic from 1993-2003]". *Sylwan* 5 (2007): 52–59.

Szlachetko, L. Dariusz. "Storczykowate (Orchidaceae) we wschodniej części Pobrzeża Kaszubskiego [Orchidaceae in the eastern part of the Kashubian Coast]". *Zeszyty Naukowe Wydz BGiO UG Biologia* 8 (1988): 99–134.Tałałaj, Iza and Emilia Brzosko. "Selfing Potential in *Epipactis palustris*, *E. helleborine* and *E. atrorubens* (Orchidaceae)". *Plant Systematics and Evolution* 276 (2008): 21–29.

Těšitelová, T., Těšitel J., Jersáková J., Rihova G., and M. A. Selosse. "Symbiotic Germination Capability of four *Epipactis* species (Orchidaceae) is Broader than Expected from Adult Ecology". *American Journal of Botany* 99 (2012): 1020–32.

Thompson, K. "Seeds and seed banks." *New Phytologist* 106 (1987): 23–34.

Thompson, K., Band S. R., and J. G. Hodgson. "Seed Size and Shape Predict Persistence in Soil." *Functional Ecology* 7 (1993): 236–41.

Tiwari, S., M. Siwakoti, B. Adhikari, and K. Subedi. *An Inventory and Assessment of Invasive Alien Plant Species of Nepal*, IUCN - The World Conservation Union, Nepal. 2005.

Tutin, T. G., V. H. Heywood, N. A. Burges, D. M. Moore, D. H. Valentine, S. M. Walters, and D. A. Webb. *Flora Europeae* 5, Cambridge: Camridge University Press, 1980.

Urbisz, Andrzej and Alina Urbisz. "Rośliny chronione południowo-zachodniej części Wyżyny Śląskiej". *Acta Biologica Silesiaca* 33, no. 50 (1998): 113–42. [Plants protected in the south-western part of the Silesian Upland. *Acta Biologica Silesiaca*]

Vakhrameeva, M. G., I. V. Tatarenko, T. I. Varlygina, G. K. Torosyan, and M. N. Zagulskii. *Orchids of Russia and Adjacent Countries (within the borders of the former USSR).* Ruggell/Liechtenstein: A.R.G. Gantner Verlag, 2008.

Veenendaal, R. L. "De bestuiving van de brede wespen orchis, *Epipactis helleborine* (L.) Crantz nader bekeken". *Nieuwsbrief Sectie Hymenoptera* NE, 31 (2010): 19–22. ["A closer look at the pollination of the broad wasp orchid, *Epipactis helleborine* (L.) Crantz". *Newsletter Section Hymenoptera*]

Vöth, W. 1982. "Blütenökologische Untersuchungen an *Epipactis atrorubens*, *E. helleborine* und *E. purpurata* in Niederösterreich. *Mitt. Bl. Arbeitskr. Heim. Orchid. Baden-Württ* 14, no. 4 (1982): 393–437. ["Flower ecological studies on *Epipactis atrorubens*, *E. helleborine* and *E. purpurata* in Lower Austria. *Mitt. Bl. Workforce Home. Orchid. Baden-Wuerttemberg*]

Waller, D. M. "Factors Influencing Seed Weight in *Impatiens capensis* (Balsaminaceae)". *American Journal of Botany* 69 (1982): 1470–75.

Whigham, D. F., J. P. O'Neill, H. N. Rasmussen, B. A. Caldwell, and M. K. McCormick. "Seed Longevity in Terrestrial Orchids – Potential for Persistent *In Situ* Seed Banks". *Biological Conservation* 129 (2006): 24–30.

Wiefelspütz, W. "Zur Verbreitung der europäischen allogamen *Epipactis*-Arten". In *Probleme der Orchideengattung Epipactis*, edited by K. Senghas, and H. Sandermann. *Jahresber. des Naturwiss. Ver. Wuppertal* 23 (1970): 38–42. ["For the distribution of the European allogamous *Epipactis* species". In Problems of the Orchid Genus

Epipactis, edited by K. Senghas, and H. Sandermann. *Annual of natural science Ver. Wuppertal*]

Wika, Stanisław. "Zagadnienia geobotaniczne środkowej części Wyżyny Krakowsko-Wieluńskiej. [Issues Geobotanical the middle of the Krakow-Wielun Upland]". *Prace Naukowe UŚ w Katowicach* 815 (1986): 1–156. ["Geobotanical issues in the central part of the Kraków-Wieluń Upland. [Issues Geobotanical the middle of the Krakow-Wielun Upland]". *Scientific Works of the University of Silesia in Katowice*]

Wittig, R., and M. Wittig. "*Epipactis helleborine* (L.) Crantz – the First (Semi)ruderal Orchid Species of Central Europe". *Feddes Repert* 118, no. 1-2 (2007): 46–50. [

Wulff, R. D. "Seed Size Variation in *Desmodium paniculatum*. I. Factors Affecting Seed Size". *Journal of Ecology* 74 (1986): 87–97.

Wyrzykiewicz-Raszewska, Małgorzata. "Struktura populacji *Epipactis palustris* (L.) Crantz – nowego apofita we florze aglomeracji Poznania". *Roczniki Akademii Rolniczej w Poznaniu seria Botanik* 4 (2001): 197–213. [Population structure of *Epipactis palustris* (L.) Crantz - a new apophyte in the flora of the Poznań agglomeration. *Annals of the Agricultural University in Poznań, series Botanics*]

Young, D. P. "Studies in the British *Epipactis* 7. Seed Dimensions and Root Diameters". *Watsonia* 5 (1962): 140–42.

Zenkert. C. A. "The Story of an Immigrant Orchid". *Hobbies* 29, no.5 (1949): 93–96.

In: Orchidaceae
Editor: Vladan Djordjević
ISBN: 978-1-68507-189-9

Chapter 8

DEVELOPMENTS IN ORCHID BREEDING THROUGH GENETIC ENGINEERING

***Ram Pal*[1,*] *and Madhurima Das*[1,2]**
[1]ICAR-National Research Center for Orchids,
Pakyong, East Sikkim, Sikkim, India
[2]Division of Plant Physiology,
ICAR-Indian Agricultural Research Institute, New Delhi, Delhi, India

ABSTRACT

With 1229 species in 184 genera, India contributes nearly 7% of the world's orchid biodiversity. Orchids are highly priced and account for 8% of the global floriculture trade, worth more than 60 billion US dollars. Large genome size, a long regeneration period, low transformation efficiency and a long life cycle have resulted in a delay in the study of orchids. Traditional introgressive breeding techniques suffer from the disadvantages of long breeding cycles (5-6 years), polyploidization of F1 generations, low availability of diverse genepools, cross-incompatibility issues, little knowledge about the genetics of trait inheritance and the requirement of considerable space

* Corresponding Author's E-mail: rampal_nrco@yahoo.com.

for growing of progenies for selection. Standardization of genetic transformation in commercial orchid genera has paved the way for successful application of various molecular techniques to achieve a rapid and desired improvement in accordance with the market demand for orchids. Gene transfer technology in orchids can change a floral trait through genetic engineering of a single or a stacked gene in far less time for the selection and regeneration of desired orchid plants and thus overcome the various limitations of traditional breeding methods. Characters like tolerance to abiotic and biotic stresses, flower color, fragrance and flowering induction are some of the general traits whose genetic transformation in orchids has been carried out successfully. Genetic engineering can target various traits of interest to producers, such as disease and pest resistance, cold hardiness in cool orchid growing regions, heat tolerance in hot ones, increased production of scent and adjustment of flowering time in keeping with market demands, thus benefitting orchid growers by reducing cultivation costs, ensuring an adequate supply of plants and allowing expansion of orchid growing into nontraditional areas.

Keywords: traditional breeding, genetic transformation, genetic engineering, traits, orchids

INTRODUCTION

Orchidaceae is one of the largest families of flowering plants in the world and consists of approximately 25,000 species in 800 genera (Arditti 1992). In addition to wild species, there are more than 150,000 artificial cultivars/hybrids. Orchids have very weak reproductive barriers, which allows members of the family to breed freely between species and between genera to form hybrids. The hybrids in turn can interbreed with other species or hybrids to form second, third, fourth, fifth, etc., generation hybrids. Some artificial intergeneric hybrids involving up to six genera have been created, e.g., *Briliandera* = *Aspasia* × *Brassia* × *Cochlioda* × *Miltonia* × *Odontoglossum* × *Oncidium* (Tan and Hew 1995). Classical plant breeding approaches in the case of orchids involve crossing between cultivated varieties, between species (interspecific hybrids) and between the genera (intergeneric hybrids). Seeds or embryos obtained from the crosses are grown to plantlets in a tissue culture laboratory, and the progenies are subsequently grown in the field. These progenies require 3-5 years to flower, upon which progenies having desirable traits are selected and multiplied for multilocational testing/ release of the variety. Thus, slow growth, a long juvenile phase, a long reproductive cycle, unknown genetics and limited genepool availability make it difficult to modify certain traits in pace with market demands. Interspecific and intergeneric hybridization with subsequent polyploidization of F_1 generations further limits their manipulation by conventional methods due to enhancement of cross-compatibility barriers, resulting in infertile capsule development or none at all. Nevertheless, conventional breeding approaches have resulted in the evolution of a large number of varieties in different cultivated genera of orchids. The lifespan of varieties in the floriculture industry is much shorter than in many other crops. Breeders are under pressure to develop new varieties in quick succession to satisfy consumer demand. Further improvement of these varieties for novel colors, plant architecture, vase life, perfume, resistance to biotic and abiotic stresses, etc., are constrained by limited genepool availability and poor understanding of the genetics of inheritance of these traits in orchids. Recent cutting-edge technologies encompassing various molecular technologies like genetic engineering, recombinant DNA technology and genetic transformation may provide a solution to the needs of growers in a shorter period of time (Teixeira da Silva et al. 2011, 2015). For successful application of transgenic technology, a standardized and high-throughput genetic transformation protocol is a pre-requsite in attempting to achieve acclimatization of orchid plants in *in vitro* regeneration and under *ex vitro* conditions. Its development has opened up a new avenue in restructuring orchid varieties for targeted traits by incorporating related genes from unrelated plants and even from non-plant organisms. For example, the first crop released for commercial cultivation in India, *Bt* cotton, drew an endotoxin-producing gene from a bacteria, *Bacillus thuringenesis*. The two major methods of gene transformation in orchids are *Agrobacterium*-mediated genetic transformation and particle gun or microprojectile bombardment, direct gene transformation being the latest addition to them. *Agrobacterium tumefaciens*-mediated genetic transformation has been reported in monocotyledonous ornamental plants, including *Oncidium*, *Dendrobium*, *Phalaenopsis*, *Anthurium* and *Iris* (Mishiba et al. 2005; Sanjaya and Chan 2007). Virus-induced gene silencing (VIGS) is a new technology just starting to be applied in orchids for functional validation of major gene silencing due to its greater efficiency and shorter turnaround time in comparison with normal mutagenesis-screening methods. Genome editing is a gene-tailoring approach in which the genome is altered by precise addition, substitution or replacement of existing DNA bases in a specific target DNA sequence (Gaj et al. 2013). A recent study reported the use of *Agrobacterium*-mediated gene transformation and application of the CRISPR/Cas9 system for editing the endogenous genes present in *Dendrobium officinale* (Kui et al. 2017).

Thus, in comparison with traditional breeding methods, genetic engineering technologies enable breeders to develop varieties at much faster rates with reduced expenditure and assured output.

GENETIC TRANSFORMATION TECHNOLOGIES

Various methods have been developed for transfer of a foreign gene into the targeted plant genome. These include *Agrobacterium*-mediated transformation, microprojectile (biolistic) bombardment, electroporation, polyethylglycol (PEG) treatment, silicon carbide-mediated transfer, microinjection, chloroplast-mediated transfer, native gene transfer, etc. However, *Agrobacterium*-mediated transformations and microparticle (biolistic) bombardments are the preferred methods for delivery of targeted DNA sequences in orchids (Kuehnle and Sugii 1992; Chia et al. 1994).

MICROPARTICLE BOMBARDMENT (BIOLISTICS)

This technology was developed by John Sanford in 1987 (Sanford et al. 1987) for delivering naked DNA into plant cells. It involves delivery of a tungsten or gold microcarrier coated with DNA into a plant cell by means of a biolistic transformation system or by gas acceleration using a particle gun. Currently, helium-driven biolistic devices along with gold microcarriers are preferred over any gun powder-driven apparatus using tungsten particles as a DNA carrier. Later devices were abandoned as they cause injury to the cells resulting in poor recovery and low transformation efficiency. The putatively transformed tissues are then tissue-cultured and selected using selection markers. In transformation of orchids, the particle bombardment method is less preferred over *Agrobacterium*-mediated transformation systems due to integration of a higher number of transformed tissues and high cost. The first successful genetic transformation using particle bombardment in orchids was achieved in the genus *Vanda* (Chia et al. 1990) and then in the genus *Dendrobium* (Kuehnle and Sugii 1992), while the first *Agrobacterium*-mediated transformation in orchids was reported eight years later (Nan et al. 1998).

The efficiency of genetic transformation of the targeted tissue in transformation through particle bombardment is determined by several physical and biological factors. Among physical factors, helium gas pressure and microcarrier size are the most important. There are contradictory views on how helium gas pressure affects transformation efficiency. Chai and Yu (2007) recommended that helium gas pressure should be optimized for each orchid species and hybrid to be transformed. Microcarrier gold particle size is another important factor found to have a great effect on gene delivery (Yang et al. 1999) and transient gene expression (Tee and Maziah 2005). Biological factors affecting transformation efficiency include the genotype, the type of tissue employed in genetic transformation and the conditions of selecting transformants. The effects of genotypes on transformation efficiency have been clearly demonstrated in studies carried out by Nan and

Kuenhle (1995), Men et al. (2003) and Chai et al. (2007). Among biological factors, the types of tissue used in genetic transformation have been found to profoundly affect transformation efficiency and transient gene expression. Protocorm-like bodies (PLBs) or calli are most frequently used as targeted tissues for transformation by this method (Nan and Kuehnle 1995; Men et al. 2003). After bombardment, putatively transformed tissues (PLBs or calli) are grown in a recovery medium devoid of any selection agent. This intermediary step between bombardment and the selection process helps the transformants to recover from the damage due to particle bombardment. The type of selection agent, concentration and duration affect the recovery of actual transformants. A higher concentration of the selection agent or prolonged exposure in the selection medium may cause death of putatively transformed explants. Conversely, a lower concentration and short exposure to the selection medium may result in selection of non-transformed explants.

Agrobacterium - Mediated Transformation

The crown gall-forming bacterium *Agrobacterium tumifaciens* has the ability to transfer a portion of T-DNA from its tumor-inducing (Ti) plasmid into the genome of a plant cell. The Ti genes of the plasmid are deleted and replaced with the DNA sequence (including selectable markers) of a desired transfer through genetic engineering (de la Riva et al. 1998). The explants are cocultivated with genetically engineered *Agrobacterium*, and putatively transformed explants are selected on a selection medium supplemented with an appropriate selection agent. The explants lacking T-DNA integration will eventually die off. *Agrobacterium*-mediated transformations have been used very successfully in dicotyledonous crop plants. However, their success in monocotyledonous plants (including orchids) is limited due their resistance to *Agrobacterium*. In dicotyledonous species, several phenolic compounds such as acetosyringone, sinapyl alcohol and coumaryl alcohol are known to induce the *vir* gene. The absence of information about these compounds in orchids has limited the application of *Agrobacterium*-mediated transformation. Nan et al. (1997) reported the presence of coniferyl alcohol as a *vir* gene inducer in PLBs of *Dendrobium* and noticed that its production and stability increased six-fold when the PLBs were cultured in light rather than in the dark. This indicated amenability of *Dendrobium* for *Agrobacterium*-mediated transformations. Nan et al. (1997) reported transformation of *Dendrobium* using *Agrobacterium tumifaciens*. This was the first report describing amenability of orchids (monocots) for genetic transformation through *Agrobacterium tumifaciens*. Soon afterwards, applicability of the *Agrobacterium*-mediated transformation protocol was extended to *Phalaenopsis* (Belarmino and Mii 2000). This list of protocols using the *Agrobacterium*-mediated transformation method in orchids has now been extended to include *Cymbidium* (Chin et al. 2007), *Dendrobium* (Nan et al. 1997; Yu et al. 2001; Men et al. 2003), *Phalaenopsis* (Mishiba et al. 2005), *Oncidium* (Liau et al. 2003), *Vanda* (Shreshtha et al. 2007) and *Catlleya* (Zhang et al. 2010).

Molecular breeders prefer *Agrobacterium*-mediated transformation over other direct methods of gene delivery because it lowers the transgene copy number, thereby causing fewer problems with transgene cosuppresion and instability (Koncz et al. 1994; Hansen et al. 1997). The single-cell transformation system also reduces problems related to the formation of mosaic plants, which more frequently arise with direct transformation methods (Enríquez-Obregón et al. 1998, 1999). Transformation efficiency in *Agrobacterium*-mediated transformation is influenced by interaction between the plant and *Agrobacterium tumifaciens*, its strain/binary vector, the choice and development stage of explants, preculture of the explant, explant wounding, the *Agrobacterium* concentration, the cocultivation medium, period length, temperature and the selection regime (Chandler and Chin 2005). Generally, the selection of transformants is done by introduction of an antibiotic or herbicide resistance genes like *hygromycin phosphotransferase II* (*hptII*), *neomycin phosphotransferase II* (*nptII*) and *phosphinothricin acetyl transferase*

(*PAT*), enabling the transformants to survive in the selection medium. There are several disadvantages of using an antibiotic and herbicide as the selective agent, e.g., possible toxicity to humans or allergenicity of the gene products in humans. The available methods for selecting transformant transgenic orchids are based on antibiotics or visual selection using other reporter genes such as green fluorescent protein (*GFP*) fusion genes, b-glucuronidase (*GUS*) and firefly *luciferase* as a screenable marker gene. To avoid the use of antibiotic or herbicide selection agents, an alternate selection system of non-plant origin has been tried in crop plants. A ferredoxin-like protein (*pflp*) was isolated from sweet pepper, and this peptide was reported to have anti-microbial activity against *Pseudomonas syringae* by delaying a hypersensitivity response in non-host plants via the release of a hairpin proteinaceous elicitor (PSS) and thus suppressing bacterial growth. Genetic transformation systems and transgene-transferred, target tissues in various orchid species were discussed in the review of Teixeira da Silva et al. (2011).

TRAITS FOR GENETIC TRANSFORMATION

In the ornamental industry, varieties with novel traits are constantly replacing existing ones to satisfy consumer demands. For this reason, varieties have a much shorter life span on the market as compared to cereals or other horticultural crops. These crops also have a smaller area of cultivation, resulting in a lower need for seed or planting materials. When gene technology is used in ornamental crops (including orchids) for commercial gains, careful consideration is required regarding the choice of crop, characteristics of the variety to be transformed and regulatory processes (Chandler and Chin 2005). General traits for modification in orchids and other ornamental crops are described here, whereas specific traits related to a particular cultivated genus have been presented elsewhere under the genus. The results of trait modification can be valuable for the producer, for the consumer or for both.

Abiotic and Biotic Stress

Improvement of traits like disease and pest resistance, tolerance to heat or cold, etc., can be of benefit to producers by reducing the cost of cultivation or its expansion into nontraditional areas. There are over 30 viruses known to infect orchids all around the world. Among them, *Cymbidium* mosaic virus (CyMV) and *Odontoglossum* ring spot virus (ORSV) are the most prevalent and economically important viruses infecting orchids worldwide (Lawson and Brannigan 1986; Zettler et al. 1990). CyMV induces floral (Yuen et al. 1979) and foliar (Dunn 1980) necrosis, while ORSV causes ring spots on leaves and color breaking on flowers (Jensen and Gold 1951). Mixed infection by both viruses can cause blossom brown necrotic streak (Thornberry and Phillippe 1964). The indicated viruses also reduce plant vigor and flower quality, which affects economic value (Okemura et al. 1983) The viruses are stable and are spread by contaminated tools and pots (Lawson and Brannigan 1986). A number of tissue culture protocols for production of virus-free plants have been developed (Ishii 1972; Albouy 1988; Lim et al. 1993; Retheesh and Bhat 2010), but keeping them virus free in cultivation is difficult due to mechanical transmission of these viruses. So far, no source for resistance or immunity is known in orchids, and hence genetic transformation for disease resistance in orchids through genetic engineering methods would indeed be of benefit to growers. Recently, a few reports on induction of disease and virus resistance in orchids through genetic transformation have been published for *Dendrobium* (Borth et al. 2004) and *Phalaenopsis* (Li et al. 2005). Tropical orchids like *Phalaenopsis* and *Dendrobium* are grown for both cut flowers and potted flowering plants. They require a high investment on energy for flower forcing during cultivation. During transportation of the finshed products (cut flowers or pot plants), they require a higher temperature than in the cold. Storage or transportation of these flowers below 100ºC causes darkening of petals and water-soaked lesions on leaves and flowers. Introduction of cold hardiness in these crops could also be useful in expanding the area of their cultivation into cooler regions and making the pot plant business more attractive. Similarly, cool-growing orchids could be transformed for increased heat tolerance, which would allow their cultivation in nontraditional areas and also reduce the energy costs of greenhouses.

Flower Color

Flower color is the most important attribute for consumers, a fact that impels breeders to develop varieties with novel colors. The main controlling genes have already been identified for the various pigment biosynthetic pathways, namely the anthocyanin (Nishihara and Nakatsuka 2011), flavonoid (Ono et al. 2006; Togami et al. 2011) and carotenoid (Sandman et al. 2006; Cazzonelli and Pogson 2010) biosynthetic pathways. Relevant knowledge about the various transcription factors controlling the pathway of anthocyanin biosynthesis (Century et al. 2008) and spatial regulation of the flavonoid biosynthetic pathway (Hichri et al. 2011) will make possible the overexpression and silencing of regulators of the respective pigment biosynthetic pathways, thereby leading to a desired phenotype. (Tanaka and Ohmiya 2008, 2009, 2010; Chandler and Brugliera 2011). Among the pigments responsible for color production, anthocyanin is the primary contributor of pink, red, blue and violet coloration via the phenyl propanoid pathway (Martin and Gerats 1993). It is a water-soluble delphinidin and malvidin (Fukuchi-Mizutani et al. 2003; Aida et al. 2000) pigment commonly accumulated in the anthocyanoplasts (Suzuki et al. 2000; Harborne and Williams 1998, 2001; Grotewold 2006; Williams and Grayer 2004). Various covalent modifications such as methylation, glycosylation and acylation are mainly responsible for the variety of anthocyanins which we see in nature. The chromophores present in the variants of anthocyanin are pelargonidin, peonidin, cyanidin and petunidin. Synthesis of anthocyanin occurs during the petal developmental stages and its regulation has been studied in the model species *Arabidopsis*, maize and petunia (Koes et al. 2005; Holton and Cornish 1995; Tsuda et al. 2004). The anthocyanin biosynthesis pathway is highly conserved and falls under the phenyl propanoid pathway. Its various genes and enzymes are *CHI, CHS, F3'5'H, F3'H, DFR* and *ANS*. The regulatory proteins for anthocyanin biosynthesis are R2R3-MYB, WDR40 and basic helix-loop transcription factors (Koes et al. 2005; Feller et al. 2011; Petroni and Tonelli 2011; Hichri et al. 2011; Hsu et al. 2015). For modification of flower color, genetic transformation technology was first used in petunia for developing orange petunias using the *dihydro-flavanol-4-reductase* gene from maize. Violet flowered 'Moon series' carnations were developed by Suntry Ltd. and Florigene Ltd. This was achieved by introducing petunia *flavonoid 3,'5'-hydroxylase* (*F3'5'H*) and petunia *dihydroflavonol 4-reductase* (*DFR*) genes into a white flowered carnation cultivar lacking the endogenous *DFR* enzyme (Nishihara and Nakatsuka 2011). Recently, Suntry Ltd. released cultivars of blue roses developed on the same line. However, as understanding of genes and gene functions is increasing, other characters like fragrance, shelf-life, plant architecture, flower characteristics and inflorescence appearance are becoming traits targeted for genetic transformation in ornamental crops (including orchids).

Flower Scent Production

Due to the cross-incompatibiity of various scented and scentless orchid species, there is a restriction in the generation of scented orchid offspring through the traditional breeding techniques available. Results of some rare successful cases of cross-breeding in orchids gave either a dilution of scent of the progeny or complete loss of scent. The biosynthetic pathways of orchid scent production and the various genes and controlling enzymatic factors are less known in orchids.

We know that floral scent is a complex process governed by many factors, and its characteristics are delineated by a mixture of low-molecular-weight volatile compounds and by various terpenoids such as phenylpropanoid, monoterpenoid, sesquiterpenoid and benzenoid compounds, as well as by fatty acid derivatives. The role of *Phalaenopsis bellina* geranyl diphosphate synthase (PbGDPS) in the fine tuning of scent production in *P. bellina* flowers is confirmed. The flower-specific *PbGDPS* gene manifests its temporal expression during flower developmental stages with maximal floral emission of monoterpenes at day five of post-anthesis, suggesting its crucial role in production of scent in orchids. The role of PbGDPS in floral scent emission is supported by findings about the enzyme in various orchid species with different scent-producing abilities (Hsiao et al. 2008). Expression of GDPS was seen in two varieties of *P. equestris*. (Hsiao et al. 2008). Research on the pathways of floral scent emission has focused on the isolation and characterization of enzymes and genes of the last steps of scent biosynthesis (terpene synthesis). Geranyl diphosphate (GDP), which is a direct precursor of scent compounds, may also play a significant role in scent production (Guterman et al. 2006; Nogues et al. 2006). Terpene synthases (TPSs) from the *P. bellina* flower and MYB transcription factors from *Phalaenopsis* exhibit a gene expression pattern similar to that of PbGDPS through microarray analysis (Hsiao et al. 2011).

Flower Induction

There are recent papers presenting the results of genome-wide analyses for the identification of genes controlling flower development in the family Orchidaceae and their species specificity (Lin et al. 2016; Huang et al. 2016; Wen et al. 2017). Isolation and functional characterization of the *FT1* gene of *Phalaenopsis aphrodite* (*PaFT1*) have been reported (Jang et al. 2015). Expression studies of *PaFT1* have been conducted in connection with the flowering behavior of this orchid. Virus-induced gene-silencing (VIGS) methods reduced the level of expression of *PaFT1*, which resulted in delayed flowering under inductive low temperature conditions (Wang et al. 2019). Similar to the *Arabidopsis* FT-FD module, *PaFT1* physically interacts with *PaFD* in *P. aphrodite* (He 2009). The leafy (*LFY*) gene (*PhapLFY*) has been reported and characterized in *P. aphrodite* (Jang 2015). Under control of the *LFY* promoter of *Arabidopsis*, it was rescued in the *Arabidopsis lfy* mutant, while *PhapLFY* overexpression in rice resulted in early heading. In *P. hybrida* (Wedding Promenade), a *CO*-like gene, *PhalCOL*, has been recently isolated (Zhang et al. 2011). It is similar in its sequence to the *CO* gene of *Arabidopsis*. Overexpression of *PhalCOL* in tobacco resulted in an early-flowering phenotype. Again, a study was done to identify various candidate genes for the control of flowering in the orchid *P. aphrodite*. For that, results of a transcriptome study in low and high ambient temperature-treated axillary buds of plants were analyzed and compared with each other (Huang et al. 2016). These results confirmed a role for *LFY*, *APETALA1* (*AP1*), *FT* and *OVEREXPRESSION OF CONSTANS 1* (*SOC1*) homologs, as well as for genes involved in the biosynthesis of GA. The transcripts of *KNOX1, R2R3-like MYB, S-adenosylmethionine synthetase, adenosine kinase 2, naringenin 3-dioxygenase* and *dihydroflavonol 4-reductase* in *Phalaenopsis* Fortune Saltzman were expressed at a higher level when grown under conditions of warm days and cool nights (28°C/21°C) as compared with daily warm temperatures (28°C/26°C) (Li et al. 2014). Low ambient temperatures promote flowering of *Dendrobium nobile* (Campos and Kerbauy 2004), and a homolog of FT was recently identified in *D. nobile* Lindl. (Li et al. 2012). In the orchid *Dendrobium* Madame Thong-In, differential gene expression studies were carried out in the SAM during its *in vitro* transition from vegetative to reproductive growth stages (Yu and Goh 2000a). A MADS-box transcription factor, namely APETALA1 (AP1), determining floral meristem identity and perianth identity in *Arabidopsis thaliana*, was isolated and characterized from *Dendrobium* Chao Praya Smile and designated as DOAP1. It is mostly expressed in inflorescence apices and various stages of flower development, and an overexpressed DOAP line was able to overcome the floral organ defects of *Arabidopsis ap1* mutants. Similarly, a study of transgenic *Dendrobium* Chao Praya Smile orchids showed DOAP1 overexpression, with earlier flowering and more rapid termination of inflorescence meristems into floral meristems as compared with the wild-type orchids (Sawettalake et al. 2017). Thus, various orthologs of *Arabidopsis* floral pathway integrater genes have been isolated and well characterized in orchids. Overexpression of *OnFT, FT, CgFT* and *DnFT* from orchids of the genera *Dendrobium*, *Oncidium* and *Cymbidium*, respectively, showed early flowering in model plants like *Arabidopsis* and *Nicotiana* (Hou and Yang 2009; Xiang et al. 2012; Li et al. 2012). Additionally, overexpression of an ortholog of *Arabidopsis* SOC1, namely DOSOC1, initiates flowering in both transgenic *Arabidopsis* and orchids of the genus *Dendrobium* (Ding et al. 2013). A MADS-box protein encoding SUPPRESSOR OF OVEREXPRESSION OF CONSTANS1 (SOC1) plays an essential role in regulation of the transition from vegetative to reproductive development in *Arabidopsis* via integration of multiple flowering signalling cues. An ortholog of SOC1, DOSOC1 was isolated from orchid species of the genus *Dendrobium* Chao Praya Smile. It was highly expressed in reproductive organs such as the floral buds, inflorescence apices and pedicels, as well as in the open flowers, and its expression increased in whole plantlets during the transition from vegetative to reproductive development. Expression of DOSOC1 was particularly seen at the stem apices in emerging floral meristems during the floral transitional stage, and its overexpression in wild-type *Arabidopsis* showed early flowering, coupled with the expression of two other flowering promoters, AGAMOUS-LIKE 24 and LEAFY. Overexpression of DOSOC1 partially complements the late-flowering phenotype of *Arabidopsis soc1-2loss-off function* mutants. It was found in one study that 35S:DOSOC1 transgenic *Dendrobium* orchid lines exhibited flowering earlier than wild-type orchids. Thus, this study reported that SOC1-like genes have an evolutionarily conserved role in promotion of flowering in the family Orchidaceae, and DOSOC1 from *Dendrobium* Chao Praya Smile serves as an important agent for genetic engineering of flowering time in the given family (Searle et al. 2006; Helliwell et al. 2006; Li et al. 2008; Ding et al. 2013).

India accounts for nearly 7% of the world's orchid biodiversity by contributing 1229 species in 184 genera. Apart from native orchids, a large number of hybrids have been developed and introduced by orchid enthusiasts and nurserymen. The international trade in floriculture products is developing rapidly and the trends are shifting fast in favor of orchids. Orchids command the highest price and are in the greatest demand. One of the main characteristics of the orchid industry is its global competitiveness. Horticulturists worldwide today favor the growing orchids not only because they are mysterious, but due to the fact that they are highly priced. They account for 8% of the global floriculture trade, worth more than 60 billion US dollars. A single high-quality quality stem of *Cymbidium* may fetch anything from Rs.130 to 1500 on the in Japanese or US market. Japan and Singapore are the main markets for orchids in Asia. The total value of orchid imports by Japan amounted to 57.4 million US dollars in 2008, making it the largest importer of orchids in the world. The main sources of imports include Thailand, Taiwan, New Zealand and Malaysia, together accounting for as much as 96.5% of the total imports in the same year. A large number of varieties of orchids belonging to the genera *Cymbidium*, *Dendrobium*, *Paphiopedilum*, *Phalaenopsis*, *Vanda*, *Oncidium* and *Zygopetalum* are commercially cultivated in various parts of the country. Details of progress made in the genetic transformation of commercially important orchids are given below.

Cymbidium

Cut *Cymbidium* flowers are prized for their beauty, elegance and long keeping quality. These are preferred flowers for gifts on special occasions or to greet elderly and important people. Apart from cut flowers, cymbidiums are also gaining popularity as flowering pot plants. The genus *Cymbidium* is composed of 44 species distributed from northwest India to China and Japan south through the Malay Archipelago to northern and eastern Australia (Du Puy and Cribb 1988). Of these, seven have been used extensively in breeding commercial cultivars and all but one (*Cymbidium sandareae*) are found in temperate regions of the world (Leonhard 1992). The species found in warmer regions bear a smaller flower and when crossed with large-flower species, their progenies are slightly warmth-tolerant and produce flowers of intermediate size. *Cymbidium* cultivars currently in cultivation are mostly advanced-generation interspecific hybrids derived from a number of species originally selected for their superior flower characters and maintained through vegetative propagation (Obara-Okeyo and Kako 1998). Vegetative growth in *Cymbidium* is promoted by long days and a day/night temperature of 30/25ºC for rapid growth and maturity of pseudobulbs (Ichihashi 1997). The variation in day and night temperatures affects flower induction. Optimal diurnal variation in temperature (20/10 and 20/14ºC) for 3 months is required for induction of flowering in large-flowered cymbidiums (Lopez and Runkle 2005). These specific temperature requirements limit their cultivation to hilly regions or make heavy expenditures necessary for heating/cooling when grown in greenhouses. Breeders have been trying to induce heat tolerance in cymbidium cultivars by conventional breeding methods. Since heat-tolerant species bear small flowers, the resulting progenies are not very attractive, either as cut flowers or as pot plants. It is impractical to increase flower size in them through introgressive breeding because the breeding cycle is too long (5-6 years). Like other orchids, cymbidiums are also affected by viral diseases that reduce their productivity and render the flowers or potted cymbidium plants unmarketable. There is no evidence indicating that any species or hybrid of the genus is immune (Jensen 1951). Disease resistance and heat tolerance can be selected as major target traits in breeding. The induction of fragrance and suppression of ethylene would further add to the crop´s attributes. A few studies have been carried out for standardization of transformation protocols in cymbidiums. The genetic transformation protocol for this group of orchids was standardized by Yang et al.(1999) using particle bombardment. In this study, before the bombardment, PLBs were cultured in liquid media for 7 days and newly formed PLBs measuring 1 mm size were chosen as target tissue for bombardment. Plasmid DNA (pKH200) carrying the *GUS-INT* and *NPTII* genes were coated onto gold particles of different size. Smaller gold particles (0.6 μm) were most effective in gene delivery. The transformed tissues were proliferated on hormone-free media for 30 days and selected in media containing 100 mg/l kanamycin against a selectable marker. The liquid culture of PLBs both before and after bombardment and the selection regime were considered important in minimizing the toxic influence of dying cells on neighboring transformed cells. Chin et al. (2007) studied *Agrobacterium*-mediated genetic transformation in three different genotypes of *Cymbidium.* They used *Agrobacterium tumifaciens* strain EHA101 harboring the binary vector pIG121Hm containing genes for *beta*-glucronidase (*GUSF*), hygromycin phosphotransferase (*hpt*) and neomycin phosphotransferase II (*nptII*) in the T-DNA region. Liquid cultured PLBs were inoculated with *Agrobacterium* and transformed PLBs were screened for hygromycin resistance. This study revealed that transformation efficiency is genetically dependent and increased by using acetosyringone (100 μl/l) in the culture media. Kanamycin is unsuitable as a selection agent. Instead, hygromycin (20 mg/l) can be used to obtain selective putative transformants of cymbidiums. Very few studies related to genetic transformation have been conducted in this group of orchids. It is one of the most commercially important genera of orchids and would be greatly benefitted by transformation for resistance for CyMV, ORSV, heat tolerance and fragrance.

Dendrobium

Dendrobium is the largest genus in the family Orchidaceae. It consists of about 1500 species, dwelling from the foothills of the Himalayas through Southeast Asia to Japan, Australia, Tasmania and the Pacific Islands. Many species, and particularly hybrids of this genus, are globally cultivated for cut flowers and potted flowering plants. Currently available hybrids have resulted from interspecies crossing, induction of polyploidy and interbreeding of species or hybrids to form several generations. The majority of hybrids were bred in Hawaii, Singapore and Thailand. Thailand is the world's largest producer and exporter of dendrobium cut flowers, accounting for more than 80% of the world's dendrobium production, projected to be worth over 100 million dollars (Hew 1994). A typical breeding cycle requires 5 to 7 years and a considaerable amount of space and labor for selecting from thousands of progenies (Xiang and Hong 2003). The main objectives in breeding of new cultivars in this genus are: i. attractive sprays; ii. long and erect to arching sprays; iii. high yields; iv. year-round flowering; and v. low bud drop and long vase life of cut flowers (Kamemoto et al. 1999). Apart from cut flowers, this genus has attracted consumers for its unique and attractive flowers, assortment of colors, year-round availability and relatively long display life. In breeding for potted flowering, pot breeders aim to obtain attractive shorter plants with upright to arching sprays, long display life, a minimum of two sprays per plant per flowering, more than one flowering a year and comparatively shorter and multiple pseudobulbs bearing green attractive leaves (Kamemoto et al. 1999). The major viruses causing significant losses to the dendrobium industry are CymMV and ORSV. No source of resistance or immunity in this genus is yet known. Genetic engineering can play a significant role in restructuring popular cultivars, thereby reducing economic losses and maintaining quality of the products. Kuehnle and Sugii (1992) used PLBs of the Jaquelyn Thomas dendrobium as target tissue for transformation of orchids. The PLBs were bombarded with microparticles coated with plasmid pGA482GG/cpPRV4 containing the *nptII* gene. Transformants were selected on a medium containing kanamycin (50-100 mg/l) and gene integration was analyzed by the polymerase chain reaction. In this study, southern blot analysis to show genomic integration of the inserted gene was not performed. This was the first study demostrating that orchids are amenable to genetic transformation through biolistic methods. The low sensitivity of kanamycin to dendrobium PLB required high doses (500 mg/l), and observation of its effect on regeneration of tissue revealed how higher doses of kanamycin can be harmful. Chia et al. (1994) proposed an alternative selection method based on expression of the introduced firefly *luciferase* gene. In this system, light-emitting cell clumps were screened and isolated using a photon-counting video camera-photomultiplier system under a high-power dissecting microscope. The screening and isolation process continued until a pure transgenic cell line was isolated. The isolated line was then induced to regenerate into a whole plant. Introduction of the *luciferase* gene as a selection marker did not find general application because of the expensive equipment required and time-consuming microscopic screening of transgenic cells. Genetic transformation in two genetically different *Dendrobium* species, *D. phalaenopsis* and *D. nobile*, using chopped calli or PLBs as target tissue was studied by Men et al. (2003). The targeted tissues were placed in a osmoticum for 1 hour before bombardment with gold particles coated with plasmid DNA containing *gus-int* and *hpt* genes. The highest percentages of transformed explants (12 and 2%) were found when the selection was carried out for two days in a selection medium supplemented with 30 mg/l hygromycin. The *gus* and *gfp* reporter systems are the ones most frequently used to monitor a transformed event in living cells. The latter is a non-destructive reporter system and is preferred over the destructive *gus* system. A successful transformation system requires good target tissues and a reporter system with a suitable promoter. Tee et al. (2003) studied target tissues and promoters of *gfp* reporter gene expression for biolistic genetic transformation in the very popular Sonia-17 *Dendrobium* cultivar. They found that pale yellow to light yellow callus with compact friable or nodal-shaped tissue morphology is the best potential target and that CaMV35S is the best promoter of *gfp* expression. Tee and Maziah (2005) evaluated both the *gfp* and the *gus* reporter systems and explored the possibility of co-bombarding multiple genes from different plasmids without tedious gene cloning and vector reconstruction. Two plasmids-p35S and pSMDFR, carrying the *gfp* and *gus* gene, respectively, were co-bombarded in two types of Sonia dendrobium calluses as suggested by Tee et al. (2003). They found that the *gfp* reporter system is a relatively faster and simpler detection method than the *gus* system, as it does not require a substrate for gene expression. Suwanaketchanatit et al. (2007) studied the microprojectile method for transformation of actively growing PLBs of the Jaqulyn Thomas dendrobium harboring *gfp* and *hpt* vector genes. Cell expressivity was identified under a fluorescent microscope during selection. A three-step screening with hygromycin in concentrations of 5 mg/dm^3, 25 mg/dm^3 and finally 30 mg/dm^3 in the selection medium was proposed instead of a higher concentration of hygromycin at a single shot to avoid chimeras and inhibition of regeneration in putative transformants with higher transformation efficiency (19.87%). The *CP* gene isolated from *Cymbidium sinense* Willed when tranferred to *Dendrobium* showed a very low level of virus accumulation, confirmed through ELISA after 4 months of inoculation (Chang et al. 2005). In monocots, genetic transformation through *Agrobacterium* requires induction of virulence (*vir*) genes in the bacterium for activation of T-DNA processing and the transfer pathway. Nan et al. (1997) for the first time demonstrated the presence of coniferyl alcohol as the *vir* gene inducer in PLBs of *Dendrobium.* The production and stability of coniferyl alcohol were enhanced almost six-fold when the PLBs were exposed to light than when maintained in the dark. Yu et al. (2001) transformed *Dendrobium* orchids with the *DOH1* antisense gene using *Agrobacterium tumifaciens* strain LBA4404 and reported that this gene caused abnormal multiple shoot development and plays some role in basic plant architecture. However, this could have been avoided by long exposure to a higher concentration of kamamycin in the selection medium. Men et al. (2003) reported a simple, rapid and efficient system for obtaining stable transgenic plants from *Dendrobium nobile* using *Agrobacterium tumifaciens* harboring *hpt* and intron-containing beta-glucuronidase (*gus-int*) genes. They activated PLBs with 100 μM acetosyrigone and co-cultivated them with *A. tumifaciens* for 2-3 days. The selection of PLBs against hygromycin (30 mg/l) and cefotaxime (250 mg/l) resulted in higher transformation efficiency (18%). Tee and Maziah (2005) standardized various parameters of particle bombardment in callusues of *Dendrobium* Sonia 17 using the green fluorescent protein (*GFP*) and b-glucuronidase (*GUS*) genes as reporter genes. The optimized parameters were acceleration pressure, gold particle size, plasmid DNA precipitation, total plasmid DNA and the ratio of plasmids co-bombarded. Chai et al. (2007) reported a novel selection agent for selecting the transgenic *Dendrobium* hybrids D. Madame Thong-In and

D. Chao Paya Smile with L-methionine-sulfoximine using the bialaphos resistance (*bar*) gene as a selectable marker. Zheng et al. (2012) developed a protocol for sonication-assisted *Agrobacterium*-mediated transformation (SAAT) of protocorm-like bodies (PLBs) in *Dendrobium* cv. 'Sanya.' They used a binary vector harboring an inverted repeat (IR) of the *ACC* gene (*IRACC*) and the neomycin phosphotransferase II gene (*NPTII*), together with an intron containing the β-glucuronidase gene (*gus int*) as a reporter gene. Treatment for 10 min with ultrasound (40 kHz) was given to the PLBs, which were and infected with an *Agrobacterium* culture containing 100 μM acetosyringone (AS). The gene transformation efficiency was calculated to be around (1.04%). After selection in neomycin and kanamycin selection agents, stable lines were generated from proliferating PLBs.

Phalaenopsis

The genus *Phalaenopsis* consists of nearly 66 species. Its habitat ranges from Sri Lanka and India in the west, to Papua New Guinea in the east, to southern China and Taiwan in the north and northern Australia in the south (Christenson 2001). The breeding of *Phalaenopsis* started as a hobby in Europe, and new hybrids were initially developed by crossing species (inter-specific) and selecting the best/elite progeny for further multiplication. In turn, these elite hybrids were further crossed with selected clones, hybrids and wild species to introduce desirable traits. Thus, they are complex hybrids that contain genetic material from several species (Gawenda et al. 2012). Breeding of *Phalaenopsis* orchids by the conventional method of sexual hybridization takes a long time and requires strenuous efforts because of their long growth and reproductive cycle (Anzai et al. 1996). Genetic transformation in orchids of the genus *Phalaenopsis* orchids using particle bombardment was first reported by Anzai et al. in 1996. In this study, PLBs derived from leaf segment culture of a *Phalaenopsis* hybrid (*Phalaenopsis* Danse × Happy Valentine) were bombarded by gold particles coated with pMSP38 and pWI-GUS. The pMSP38 contained a bialaphos- resistant gene (*bar*) driven by the CaMV35S promotor and nopaline synthase (*nos*) terminator, whereas pWI-GUS contained the *GUS* gene driven by the CaMV35S promoter and 35S terminator. The bombarded PLBs were cultured on a medium devoid of bialaphos for 2 months. Thereafter, proliferated PLBs were screened for bialaphos resistance on culture media conatining 1 mg/l bialaphos. The survived PLBs were further screened on a medium containing 5 mg/l bialaphos. Only seven bialaphos-resistant plantlets were obtained, of which only one was *GUS*-positive. Belarmino and Mii (2000) reported a protocol for susscessful production of transgenic plantlets in phalaenopsid orchids using cell clumps as explants. Chai et al. (2002) reported regeneration of genetically transformed PLBs of phalaenopsid orchids. Transversly bisected ones are better for proliferation of PLBs. Regenerated PLBs when subjected to a selection cycle with a long selection period and low hygromycin concentration (1.5 mg/L) yielded stable transformants. You et al. (2003) used the *pflp* gene in *Agrobacterium*-mediated tranformation and *E. caratovora* as a selection agent. Enhanced resistance to *E. caravora* in trangenic PLBs was observed. In their opinion, the *pflp* gene could be a useful selection candidate and bacterial resistance gene in genetic engineering strategies. Liao et al. (2004) introduced cymbidium mosaic resistance in *Phalaenopsis* orchids using a DNA cassette containing a CymMV coat protein (CP) and cDNA. A NOS terminator placed downstream from an ubiquitin promotor was transformed into a *Phalaenopsis* orchid by particle bombardment. Resistance in transgenic plants was found to be RNA-mediated through a post-transcriptional gene-silencing mechanism (PTGS). Chan et al. (2005) genetically engineered a *Phalaenopsis* orchid resistant to cymbidium mosaic virus (CymMV) and soft rot disease caused by *Erwinia caratovora*. Gene stacking has been used against two of the most dreaded disesases of orchids. This was done by tranforming PLBs with CymMV coat protein cDNA (*CP*) and then retransferring with sweet pepper-like protein cDNA (*Pflp*) by *Agrobacterium tumifaciens*. Transgenic *Phalaenopsis* orchids resistant to both viral and bacterial infection were then established using antibiotic and non-antibiotic selection agents. Mishiba et al. (2005) used early-stage protocorms of a cross combination of heterozgous cultivars of *Phalaenopsis* orchids for *Agrobacterium*-mediated tranformations. The transformation efficiency was 1.3-1.9%. Sjahril and Mii (2006) used embryonic cells for *Agrobacterium*-mediated transformation of the washabi defensin gene, an antibacterial protien-encoding gene (*WjAMP-1*) isolated from *Wasabia japonica* (Sjahril et al. 2001) in *Phalanoepsis* orchids in order to induce resistance to soft rot caused by *Erwinia caratovora*. The transgenic plants showed strong resistance to *E. caratovora*. Over-expression of the gene further enhanced the resistance. Shrestha et al. (2007) reported an *Agrobacterium*-mediated transformation protocol for *Phalaenopsis amabilis* using *gus* genes and demonstrated usefulness of the *BP/KANT 1* gene, responsible for shoot apical meristem development in *Arabidopsis thaliana*, as a visible selection marker by vitue of producing many shoots. In order to induce cold resistance in embryogenic callus of *Phalaenopsis amabilis* through genetic transformation, Qin et al. (2011) infected and cocultivated calli derived from a leaf culture with *Agrobacterium tumefaciens* strain LBA4404, which harbors a plant-cloning vector carrying a lipid transfer protein (*LTP*) encoding a gene cloned from cold-tolerant Brazilian upland rice cv. IAPAR 9. Expression of the rice cold-inducible *LTP* gene in transgenic *P. amabilis* improved its adaptive responses to cold stress. Analysis of transgenic plants indicated that enhanced cold tolerance was most likely due to increased accumulation of proline, total soluble sugars and superoxide dismutase, decreased accumulation of malondialdehyde and better maintainance of electrolytes within the membrane as compared with the controls.

Vanda

The genus *Vanda* consists of about 50 species, distributed from the Himalayas and southern China to Sri Lanka, and in Southeast Asia to Papua New Guinea and northern Australia (Soon 2005). This genus breeds well between species and with many other genera. Several man-made genera like *Ascocenda* (*Ascocentrum* × *Vanda*), *Vascostylis* (*Vanda* × *Ascocentrum* × *Rhyncostylis*), *Aranda* (*Arachinis* × *Vanda*), *Mokara* (*Arachinis* × *Vanda* × *Ascocentrum*), *Renantanda* (*Renananthera* × *Vanda*), *Rhyncovanda* (*Rhyncostylis* × *Vanda*), etc., have been developed. Breeding with such huge diverse and unrelated genera has resulted in wide variability in plant architecture, flower forms, size and shapes of flowers, flowering habits, etc. The cultivars of this group bear stunning colors, pure yellow, orange, scarlet and deep purple, or delicately marked with stripes or spots. Inflorescence longevity in these orchids varies from 6 to 8 weeks and blooming occurs six or more times a year (Motes 2004). The majority of vandas are cultivated in Thailand and Hawaii. These are tropical plants that like to grow in a temperature range of 15 to 32ºC and under light of higher intensity. Though considerable variability exists in cultivated varieties, the acquisition of additional characters like resistance to pests and disease, cold tolerance and tolerance of low light intensity through genetic engineering is desirable (Shreshtha et al. 2007). Genetic transformation in *Vanda* was first briefly described by Chia et al. (1990). Recently, Shreshtha et al. (2007) reported a sonication-assisted *Agrobacterium*-mediated transformation protocol for the 'Tokya Blue' vanda. They used *Agrobacterium tumefaciens* strain EHA101 harboring the binary vector plasmid pIG121Hm containing a hygromycin phosphotransferase gene (*hpt)* and a neomycin phosphotranferase II gene (*nptII*) as selectable markers and an intron-containing β-glucuronidase (*gus*) reporter gene. They concluded that a long period of infection (4 h), wounding of PLBs and sonification (5 min) increase the tranformation efficiency. The transformation efficiency was 17 plants per g PLBs.

Cattleya

The genus *Cattleya* consists of 48 species, distributed at middle elevations in Central and South America, and more than half of the species described under this genus are found in Brazil. Genetic transformation in *Cattleya* was first reported by Knapp et al. (2000). Protocorms of the *Cattleya Georgianna* × self were bombarded with gold particles (1 μm) coated with plasmid DNA (pDPG165) containing the *bar* gene, a promoter (CaMV35S) and a terminator (Tr7). Transformants were selected following a two-step selection process after 3 weeks on a medium containing an alanine derivative and 1 mg/l bialaphos, followed by 8 weeks on a medium containing 3 mg/l bialaphos. The *bar* gene confers resistance to a broad spectrum of herbicides such as Phosphinothrin (PPT). For selection of transformants, *Agrobacterium*-mediated genetic transformation has several advantages over other direct methods of gene delivery into the host cell. In *Cattleya*, an *Agrobacterium*-mediated transformation protocol was standardized by Zhang et al. (2010). They cocultivated PLBs of the CM2450 *Cattleya* with *Agrobacterium tumifaciens* strain EHA101, carrying plasmid PIG121-Hm, which contains genes coding for *nptII*, *hpt* and *GUS*, or plasmid pBBRacdS, harboring the same genes plus a gene coding for 1-aminocyclopropane-1-carboxylase (ACC) deaminase. This gene is reported to increase transformation efficiency by decreasing the level of ethylene in target tissue. Transformation efficiency was found to be influenced by the bacterial concentration and cocultivation period. An *Agrobacterium* cell concentration of 1:10 ($OD_{600} = 0.6$) and cocultivation for 3 hours was optimum for obtaining the highest number of transformants (6.3/g) in this cattleya. Harboring pBBRacdS, a plasmid with a gene coding for 1-aminocyclopropane-1-carboxylase (ACC) deaminase, had no significant effect on transformation efficiency. Zheng et al. (2010) produced plants resistant to *Odontoglossum* ring spot virus (ORSV) using a gene coding for ORSV replicase in T-DNA of a plasmid (pSMAHdN627) carried by *Agrobacterium tumifaciens* strain EHA105. They found that transformation efficiency is highest (11/1 g PLBs) when the PLBs are pre-cultured for 3 hours in the presence of acetosyringone (AS) and cocultivated in an *Agrobacterium* suspension (1:10 OD_{600} = 0.6) for 3 days in the dark.

Oncidium

Liau et al. (2003) extended the *Agrobacterium*-mediated genetic transformation technology to the genus *Oncidium.* They used *Agrobacterium tumefaciens* strains to which plasmid (pCAMBIA1301) containing *hptII* and *gusA* genes were introduced by electroporation. In order to standardize the protocol, the PLBs of the Sharry Babay oncidium were chosen as targeted tissues, and the sensitivity of PLBs to hygromycin, age of PLBs and effectiveness of strains (EHA105 and LBA4404) were studied. The lowest hygromycin concentration required for screening of was found to be 5 mg/l and the optimal age of PLBs for transformation was 45 days. The transformation efficiency was not affected by strains. However, the shoot formation rate of EHA105-transformed PLBs was higher than in LBA4404-transformed PLBs. Raffeiner et al. (2009) used the *nptII* gene under the control of the nopalin synthase gene (*nos*) was used as a selectable marker. The lowest concentration of kanamycin for selection of putative transformants was found to be 300 mg/l. Preculture of PLBs on a solid medium for 3 days and then infecting them with a bacterial suspension having $OD_{600} = 1.0$ showed the highest percentage of *GUS*-positive PLBs. The protocol was used to tranform the Sweet Sugar oncidium with the *ert1-1* gene-known for controlling senesence in *Arabidopsis thaliana* under the influence of a flower-specific promoter (*fpb1*) from a petunia hybrid. Ferredoxin-like protein (*pflp*) isolated from sweet pepper showed antimicrobial activity against *Pseudomonas syringae* (Lin et al. 1997). The *Pflp* gene, encoding for sweet pepper ferredoxin-like protein, has been used to induce resistance against *Xanthomonas oryzae* pv oryzeae, which causes blight disease in rice (Tang et al. 2001). Liau et al. (2003) successfully integrated the *pflp* gene into the *Oncidium* genome via *Agrobacterium*-mediated transformation to confer resistance against *Erwinia caratovora*, a pathogen that causes soft rot disease in orchids of the given genus. The resistance of transformants was confirmed by placing wounded leaves in a bacterial suspension culture and injecting the bacterial suspension into shoots. The transformants showed resistance to *Erwinia caratovora.*

Thiruvengadam et al. (2011) used the phosphomanose-isomerase gene (*pmi*) as a selectable marker in the Grower Ramsay oncidium. Plants with this gene are able to convert manose-6-phosphate to fructose-6-phosphate and continue their growth in manose media, whereas plants lacking this gene get eliminated due to death by starvation on a manose medium. Genetically transformed plants are selected in this way. A selection medium containing 25 g/l manose and 5 g/l sucrose were found to be suitable for selection of transformants. Li et al. (2005) reported that pretreatment of PLBs of the Sharry Babay 'OM8' oncidium with 0.5 M sucrose for 2 hours increased single cell embryogenesis and *GUS* expression 14.8-fold. Yee et al. (2008) optimized the protocol for co-transfer of common marker genes (*gfp* and *hptII*) and a flower color gene (*CHS*) to the Sharry Baby oncidium using a biolistic gun. They demonstrated that hygromycin in a concentration of 5 g/l is a good seletion agent for effective transformation. For greater tranformation efficiency, 4-week old PLBs of 5-6 mm size and preinoculation of a fresh medium one day before bombardment with 1 g DNA prepared on calcium was found to be important in biolistic tranformation of the given oncidium. Wang et al. (2016) reported using RNAi of the b-ring carotene hydroylase gene (*BCH2*), (which is involved in carotenoid biosynthesis, for pigment formation in a transgenic oncidium with the aid of *Agrobacterium tumefaciens*, thereby changing the color from bright yellow to light yellow and white-yellow. Thus, the authors of that study successfully modified flower color in an oncidium by downregulation of genes related to carotenoid biosynthesis and further facilitated exploitation of the promoter regions of *BCH2* genes for molecular breeding of oncidiums.

CONSTRAINTS

The study of orchids is hampered to a large extent by their large genome size, long regeneration period, low transformation efficiency and long life cycle. Establishing efficient and reproducible gene transformation systems for molecular breeding in orchids is a crucial challenge. Particle bombardment in orchids is successful, but with low transformation and regeneration rates. In orchids, the most challenging and intriguing studies are ones devoted to the functional validation of genes involved in reproductive growth stages, as it takes nearly 2-3 years for the transition from vegetative to reproductive stages. Moreover, the feasibility of trait engineering is not only dependent on a single gene, but may require manipulation of metabolic pathways, which is associated with multiple genes. Thus, a common technical problem is to achieve the efficient functioning of these multiple transgenes.

FUTURE SCOPE

Through genetic transformation of orchids, which are the most highly valued floricultural crop in terms of market demand, various traits such as transition to the flowering state from the vegetative state, flower color, the fragrance of flowers and their architechture can potentially be altered within a short period of time, thereby averting traditional breeding strategies. Methods of genetic transformation have been standardized in commercially available orchid genera. The reduction in time and economic savings recently obtained by using *pflp* and L-methionine sulfoximine as selection agents may help to facilitate studies related to orchid genes. Again, an important method in studying the functions of genes is to eliminate their expression through loss-of-function mutants. Recently employed in orchids, virus-induced gene silencing (VIGS) is a viable alternative to loss-of-function assays that takes much less time. Other advantages of VIGS are that it can knock down target dsRNA transcribed from RNA at different locations in the genome and thus silence all genes simultaneously, testing can be conducted within a month after virus inoculation and the technique can be applied at any stage of plant development. Challenges for future research on orchids would be to identify additional genes controlling flowering time, symmetry, flower morphogenesis and the role of transposable elements. Additionally, embryogenesis, disease resistance, genetic diversity and mycoheterotrophic interactions are areas that will benefit from using the tools of molecular biology developed in studying orchid flower morphogenesis.

CONCLUSION

Conventional breeding can be used to transmit useful traits into commercial varieties. Development of novel varieties is a cumbersome process in orchids, especially due to slow growth and cross-incompatibility. In the future, genetic engineering is likely to play a very significant role in development of new varieties. The discovery, isolation and identification of orchid floral genes and a better understanding of genomics and functional genomics will lead to greater applicability of genetic transformation. Further, RNAi, VIGS and genome editing technologies like CRISPER/Cas9 have helped us to understand the molecular regulatory mechanisms governing floral induction/ development, scent production and color presentation in orchids to satisfy market demands. Gene transfer technology in orchids has the potential to change a particular desired floral trait via genetic engineering of a single or a stacked gene and thus overcome the various limitations of traditional breeding methods. New molecular techniques enable breeders to achieve the selection and regeneration of desired orchid plants in far less time. Farther on in the future, these genetic transformation technologies can be applied in areas of high-priorty importance in orchids such as generation of tolerance towards various abiotic and biotic stresses, especially in combating viral diseases [as reported to date in *Dendrobium* (Chang et al. 2005), *Phalaenopsis* (Chan et al. 2005; Zheng 2008), *Cymbidium* (Shi et al. 2011) and *Cattleya* (Zhang et al. 2010b)]; delaying flower scenescence; and increasing the post-harvest shelf life of flowers. The CRISPER/Cas9 genome-editing technique can be successfully utilized in genetic transformation of orchids for precise modification of the genome, ultimately producing novel genotypes.

REFERENCES

Aida, Ryutaro, Kumi Yoshida, Tadao Kondo, Sanae Kishimoto, and Michio Shibata. "Copigmentation gives bluer flowers on transgenic torenia plants with the antisense dihydroflavonol-4-reductase gene." *Plant Science* 160, no. 1 (2000): 49–56.

Albouy, J., C. Flouzat, C. Kusiak, and Maurice Tronchet. "Eradication of orchid viruses by chemotherapy from in-vitro cultures of *Cymbidium*." In *VII International Symposium on Virus Diseases of Ornamental Plants* 234, pp. 413–20. 1988.

Anzai, Hiroyuki, Yukiko Ishii, Makoto Shichinohe, Kazuko Katsumata, Chuhei Nojiri, Hiromichi Morikawa, and Michio Tanaka. "Transformation of *Phalaenopsis* by particle bombardment." *Plant Tissue Culture Letters* 13, no. 3 (1996): 265–72.

Arditti, Joseph. *Fundamentals of orchid biology*. John Wiley & Sons, 1992.

Belarmino, M. M., and M. Mii. "*Agrobacterium*-mediated genetic transformation of a *Phalaenopsis* orchid." *Plant Cell Reports* 19, no. 5 (2000): 435–42.

Borth, W. B., K. Barry, K. Obsuwan, M. Q. Xu, R. W. Liu, A. R. Kuehnle, and J. S. Hu. "Movement of *Cymbidium* mosaic virus and transgenic resistance in *Dendrobium* orchids." In *XI International Symposium on Virus Diseases of Ornamental Plants 722*, pp. 137–146, 2004.

Campos, Kátia Oliveira, and Gilberto Barbante Kerbauy. "Thermoperiodic effect on flowering and endogenous hormonal status in *Dendrobium* (Orchidaceae)." *Journal of Plant Physiology* 161, no. 12 (2004): 1385–87.

Cazzonelli, Christopher I., and Barry J. Pogson. "Source to sink: regulation of carotenoid biosynthesis in plants." *Trends in Plant Science* 15, no. 5 (2010): 266–74.

Century, Karen, T. Lynne Reuber, and Oliver J. Ratcliffe. "Regulating the regulators: the future prospects for transcription-factor-based agricultural biotechnology products." *Plant Physiology* 147, no. 1 (2008): 20–29.

Chai, Davina, and Hao Yu. "Recent advances in transgenic orchid production." *Orchid Sci. Biotechnol.* 1, no. 2 (2007): 34–39.

Chai, Davina, Sook May Lee, Jia Hui Ng, and Hao Yu. "L-methionine sulfoximine as a novel selection agent for genetic transformation of orchids." *Journal of Biotechnology* 131, no. 4 (2007): 466–72.

Chai, M. L., C. J. Xu, K. K. Senthil, J. Y. Kim, and D. H. Kim. "Stable transformation of protocorm-like bodies in *Phalaenopsis* orchid mediated by *Agrobacterium tumefaciens*." *Scientia Horticulturae* 96, no. 1–4 (2002): 213–24.

Chan, Yuan-Li, Kuang-Hung Lin, Li-Jen Liao, Wen-Huei Chen, and Ming-Tsair Chan. "Gene stacking in *Phalaenopsis* orchid enhances dual tolerance to pathogen attack." *Transgenic Research* 14, no. 3 (2005): 279–88.

Chandler, Stephen F., and Chin-Yi Lu. "Biotechnology in ornamental horticulture." *In Vitro Cellular & Developmental Biology-Plant* 41, no. 5 (2005): 591–601.

Chandler, Stephen F., and Filippa Brugliera. "Genetic modification in floriculture." *Biotechnology Letters* 33, no. 2 (2011): 207–14.

Chang, Chen, Ying-Chun Chen, Yau-Heiu Hsu, Jian-Tai Wu, Chung-Chi Hu, Wei-Chin Chang, and Na-Sheng Lin. "Transgenic resistance to *Cymbidium* mosaic virus in *Dendrobium* expressing the viral capsid protein gene." *Transgenic Research* 14, no. 1 (2005): 41–46.

Chia, T. F., Y. S. Chan, and N. H. Chua. "Large-scale screening of *Cymbidium* mosaic and *Odontoglossum* mosaic ringspot viruses in cultivated orchids by nucleic acid spot hybridization." In *Proceedings of the 13th World Orchid Conference. 13th WOC Proceedings Trust,* Auckland, New Zealand, p. 284, 1990.

Chia, Tet-Fatt, Yang-Sun Chan, and Nam-Hai Chua. "The Firefly Luciferase Gene as a Non-Invasive Reporter for *Dendrobium* Transformation." *The Plant Journal: For Cell and Molecular Biology* 6, no. 3 (1994): 441–46.

Chin, D. P., M. Mii, and K. I. Mishiba. "Production of transgenic *Phalaenopsis* plants by introducing glutathione S-transferase gene into protocorms at an early stage after germination." In *XXII International Eucarpia Symposium, Section Ornamentals, Breeding for Beauty-Part II* 743, pp. 101–105, 2006.

Chin, Dong Poh, Kei-ichiro Mishiba, and Masahiro Mii. "*Agrobacterium*-mediated transformation of protocorm-like bodies in *Cymbidium*." *Plant Cell Reports* 26, no. 6 (2007): 735–43.

Christenson E. A. *Phalaenopsis*. pp. 330, Portland, OR, USA: Timber Press, 2001.

Da Silva, Jaime A. Teixeira, Dong Poh Chin, Pham Thanh Van, and Masahiro Mii. "Transgenic orchids." *Scientia Horticulturae* 130, no. 4 (2011): 673–80.

Da Silva, Jaime A. Teixeira, Jean Carlos Cardoso, Judit Dobránszki, and Songjun Zeng. "*Dendrobium* micropropagation: a review." *Plant Cell Reports* 34, no. 5 (2015): 671–704.

De La Riva, Gustavo A., Joel González-Cabrera, Roberto Vázquez-Padrón, and Camilo Ayra-Pardo. "*Agrobacterium tumefaciens*: a natural tool for plant transformation." *Electronic Journal of Biotechnology* 1, no. 3 (1998): 24–25.

Ding, Lihua, Yanwen Wang, and Hao Yu. "Overexpression of DOSOC1, an ortholog of *Arabidopsis* SOC1, promotes flowering in the orchid *Dendrobium* Chao Parya Smile." *Plant and Cell Physiology* 54, no. 4 (2013): 595–608.

Du, Puy., and P. Cribb. *The classification of Cymbidium.* pp. 50–194. Oregon, USA: Timber Press, 1988.

Dunn, E. "Virus disease of *Dendrobium*." *Third Annual Ornamental Short Course. Proc. Hawaii Inst. Trop. Agric. Hum. Resour. Res. Ext. Ser 7* (1980): 77–81.

Enríquez-Obregón, Gil A., Dmitri L. Prieto-Samsónov, A. Gustavo, Marlene Pérez, Guillermo Selman-Housein, and Roberto I. Vázquez-Padrón. "*Agrobacterium*-mediated Japonica rice transformation: a procedure assisted by an antinecrotic treatment." *Plant Cell, Tissue and Organ Culture* 59, no. 3 (1999): 159–68.

Enríquez-Obregón, Gil A., Roberto I. Vázquez-Padrón, Dmitri L. Prieto-Samsonov, A. Gustavo, and Guillermo Selman-Housein. "Herbicide-resistant sugarcane (*Saccharum officinarum* L.) plants by *Agrobacterium*-mediated transformation." *Planta* 206, no. 1 (1998): 20–27.

Feller, Antje, Katja Machemer, Edward L. Braun, and Erich Grotewold. "Evolutionary and comparative analysis of MYB and bHLH plant transcription factors." *The Plant Journal* 66, no. 1 (2011): 94–116.

Fukuchi-Mizutani, Masako, Hiroaki Okuhara, Yuko Fukui, Masahiro Nakao, Yukihisa Katsumoto, Keiko Yonekura-Sakakibara, Takaaki Kusumi, Toshiharu Hase, and Yoshikazu Tanaka. "Biochemical and molecular characterization of a novel UDP-glucose: anthocyanin 3′-O-glucosyltransferase, a key enzyme for blue anthocyanin biosynthesis, from gentian." *Plant Physiology* 132, no. 3 (2003): 1652–63.

Gaj, Thomas, Charles A. Gersbach, and Carlos F. Barbas III. "ZFN, TALEN, and CRISPR/Cas-based methods for genome engineering." *Trends in Biotechnology* 31, no. 7 (2013): 397–405.

Gawenda, I., A. Schröder-Lorenz, and T. Debener. "Markers for ornamental traits in *Phalaenopsis* orchids: population structure, linkage disequilibrium and association mapping." *Molecular Breeding* 30, no. 1 (2012): 305–16.

Grotewold, Erich. "The genetics and biochemistry of floral pigments." *Annual Review of Plant Biology* 57 (2006): 761–80.

Guterman, Inna, Tania Masci, Xinlu Chen, Florence Negre, Eran Pichersky, Natalia Dudareva, David Weiss, and Alexander Vainstein. "Generation of phenylpropanoid pathway-derived volatiles in transgenic plants: rose alcohol acetyltransferase produces phenylethyl acetate and benzyl acetate in petunia flowers." *Plant Molecular Biology* 60, no. 4 (2006): 555–63.

Hansen, Geneviève, Raymond D. Shillito, and Mary-Dell Chilton. "T-strand integration in maize protoplasts after codelivery of a T-DNA substrate and virulence genes." *Proceedings of the National Academy of Sciences* 94, no. 21 (1997): 11726–30.

Harborne, Jeffrey B., and C. A. Williams. “Anthocyanins and other flavonoids.” *Natural Product Reports* 15, no. 6 (1998): 631–52.

Harborne, Jeffrey B., and Christine A. Williams. “Anthocyanins and other flavonoids.” *Natural Product Reports* 18, no. 3 (2001): 310–33.

He, Yuehui. “Control of the Transition to Flowering by Chromatin Modifications.” *Molecular Plant* 2, no. 4 (2009): 554–64.

Helliwell, Chris A., Craig C. Wood, Masumi Robertson, W. James Peacock, and Elizabeth S. Dennis. “The *Arabidopsis* FLC protein interacts directly in vivo with SOC1 and FT chromatin and is part of a high-molecular-weight protein complex.” *The Plant Journal* 46, no. 2 (2006): 183–92.

Hew, C. S. “Orchid cut flower production in ASEAN countries.” In *Orchid biology: Reviews and prospectives*, Vol VI, edited by J. Arditti, 363–401. New York: Wiley, 1994.

Hichri, Imène, François Barrieu, Jochen Bogs, Christian Kappel, Serge Delrot, and Virginie Lauvergeat. “Recent advances in the transcriptional regulation of the flavonoid biosynthetic pathway.” *Journal of Experimental Botany* 62, no. 8 (2011): 2465–83.

Holton, Timothy A., and Edwina C. Cornish. “Genetics and biochemistry of anthocyanin biosynthesis.” *The Plant Cell* 7, no. 7 (1995): 1071.

Hou, Cheng-Jing, and Chang-Hsien Yang. “Functional analysis of FT and TFL1 orthologs from orchid (*Oncidium* Gower Ramsey) that regulate the vegetative to reproductive transition.” *Plant and Cell Physiology* 50, no. 8 (2009): 1544–57.

Hsiao, Yu-Yun, Mei-Fen Jeng, Wen-Chieh Tsai, Yu-Chen Chuang, Chia-Ying Li, Tian-Shung Wu, Chang-Sheng Kuoh, Wen-Huei Chen, and Hong-Hwa Chen. “A novel homodimeric geranyl diphosphate synthase from the orchid *Phalaenopsis bellina* lacking a DD (X) 2–4D motif.” *The Plant Journal* 55, no. 5 (2008): 719–33.

Hsiao, Yu-Yun, Zhao-Jun Pan, Chia-Chi Hsu, Ya-Ping Yang, Yi-Chin Hsu, Yu-Chen Chuang, Hsing-Hui Shih, Wen-Huei Chen, Wen-Chieh Tsai, and Hong-Hwa Chen. “Research on orchid biology and biotechnology.” *Plant and Cell Physiology* 52, no. 9 (2011): 1467–86.

Hsu, Chia-Chi, You-Yi Chen, Wen-Chieh Tsai, Wen-Huei Chen, and Hong-Hwa Chen. "Three R2R3-MYB transcription factors regulate distinct floral pigmentation patterning in *Phalaenopsis* spp." *Plant Physiology* 168, no. 1 (2015): 175–91.

Huang, Jian-Zhi, Chih-Peng Lin, Ting-Chi Cheng, Ya-Wen Huang, Yi-Jung Tsai, Shu-Yun Cheng, Yi-Wen Chen et al. "The genome and transcriptome of *Phalaenopsis* yield insights into floral organ development and flowering regulation." *Peer J* 4 (2016): e2017.

Ichihashi, Syoichi. "Orchid production and research in Japan." In *Orchid Biology*, pp. 171–212. Springer, Dordrecht, 1997.

Ishii, M. "Partial elimination of virus from doubly infected orchids by meristem explant culture." In *III International Symposium on Virus Diseases of Ornamental Plants* 36, pp. 229–34, 1972.

Jang, Seonghoe. "Functional characterization of PhapLEAFY, a FLORICAULA/LEAFY ortholog in *Phalaenopsis aphrodite*." *Plant and Cell Physiology* 56, no. 11 (2015): 2234–47.

Jensen, D. D. "Mosaic or black streak disease of *Cymbidium* Orchids." *Phytopathology* 41, no. 5 (1951).

Jensen, D. D., and A. H. Gold. "A virus ring spot of *Odontoglossum* Orchid: symptoms, transmission, and electron microscopy." *Phytopathology* 41, no. 7 (1951).

Kamemoto, Haruyuki, Teresita D. Amore, and Adelheid R. Kuehnle. "*Breeding Dendrobium orchids in Hawaii*." University of Hawaii Press, 1999.

Knapp, J. E., A. P. Kausch, and J. M. Chandlee. "Transformation of three genera of orchid using the bar gene as a selectable marker." *Plant Cell Reports* 19, no. 9 (2000): 893–98.

Koes, Ronald, Walter Verweij, and Francesca Quattrocchio. "Flavonoids: a colorful model for the regulation and evolution of biochemical pathways." *Trends in Plant Science* 10, no. 5 (2005): 236–42.

Koncz, C., Németh, K., Redei, G. P. and Scell, J. (1994). In *Homologous recombination and gene silencing in plants*, edited by J. Paszkowski, 167–189. Kluwer, Dordrecht, The Netherlands.

Kuehnle, Adelheid R., and Nellie Sugii. "Transformation of *Dendrobium* orchid using particle bombardment of protocorms." *Plant Cell Reports* 11, no. 9 (1992): 484–88.

Kui, Ling, Haitao Chen, Weixiong Zhang, Simei He, Zijun Xiong, Yesheng Zhang, Liang Yan et al. "Corrigendum: building a genetic manipulation tool box for orchid biology: identification of constitutive promoters and application of CRISPR/Cas9 in the orchid, *Dendrobium officinale*." *Frontiers in Plant Science* 8 (2017): 664.

Lawson, R. H., and M. Brannigan. "*Handbook on orchid pests and diseases*." Amer. Orchid Soc., West Palm Beach, Fla (1986): 2–49.

Leonhardt, K. W. "*Cymbidium* species and hybrids for the Asian tropics." *Proceedings of the Fourth Asia Pacific Orchid Conference* (APOC 4), Chiang Mai, Thailand, 25-26 January, 1992, pp. 35–40.

Li, D. M., F. B. Lu, G. F. Zhu, Y. B. Sun, Y. C. Xu, M. D. Jiang, J. W. Liu, and Z. Wang. "Identification of warm day and cool night conditions induced flowering-related genes in a *Phalaenopsis* orchid hybrid by suppression subtractive hybridization." *Genetics and Molecular Research* 13, no. 3 (2014): 7037–51.

Li, Dan, Chang Liu, Lisha Shen, Yang Wu, Hongyan Chen, Masumi Robertson, Chris A. Helliwell, Toshiro Ito, Elliot Meyerowitz, and Hao Yu. "A repressor complex governs the integration of flowering signals in *Arabidopsis*." *Developmental Cell* 15, no. 1 (2008): 110–20.

Li, Ruihong, Aike Wang, Shulan Sun, Shan Liang, Xiaojing Wang, Qingsheng Ye, and Hongqing Li. "Functional characterization of FT and MFT ortholog genes in orchid (*Dendrobium nobile* Lindl) that regulate the vegetative to reproductive transition in *Arabidopsis*." *Plant Cell, Tissue and Organ Culture (PCTOC)* 111, no. 2 (2012): 143–51.

Li, Shu-Hua, Chang-Sheng Kuoh, Yau-Huang Chen, Hong-Hwa Chen, and Wen-Huei Chen. "Osmotic sucrose enhancement of single-cell embryogenesis and transformation efficiency in *Oncidium*." *Plant Cell, Tissue and Organ Culture* 81, no. 2 (2005): 183–92.

Liao, Li-Jen, I-Chun Pan, Yuan-Li Chan, Yau-Heiu Hsu, Wen-Huei Chen, and Ming-Tsair Chan. "Transgene silencing in *Phalaenopsis* expressing the coat protein of *Cymbidium* Mosaic Virus is a manifestation of RNA-mediated resistance." *Molecular Breeding* 13, no. 3 (2004): 229–42.

Liau, C-H., S-J. You, V. Prasad, H-H. Hsiao, J-C. Lu, N-S. Yang, and M-T. Chan. "*Agrobacterium tumefaciens*-mediated transformation of an *Oncidium* orchid." *Plant Cell Reports* 21, no. 10 (2003): 993–98.

Liau, Chia-Hui, Jian-Cheng Lu, Venkatesh Prasad, Hsin-hao Hsiao, Su-Juan You, Jent-turn Lee, Ning-Sun Yang et al. "The sweet pepper ferredoxin-like protein (pflp) conferred resistance against soft rot disease in *Oncidium* orchid." *Transgenic Research* 12, no. 3 (2003): 329–36.

Lim, S. T., S. M. Wong, and C. J. Goh. "Elimination of cymbidium mosaic virus and odontoglossum ringspot virus from orchids by meristem culture and thin section culture with chemotherapy." *Annals of Applied Biology* 122, no. 2 (1993): 289–97.

Lin, Choun-Sea, Chen-Tran Hsu, De-Chih Liao, Wan-Jung Chang, Ming-Lun Chou, Yao-Ting Huang, Jeremy J. W. Chen, Swee-Suak Ko, Ming-Tsair Chan, and Ming-Che Shih. 2016. "Transcriptome-Wide Analysis of the MADS-Box Gene Family in the Orchid *Erycina pusilla*." *Plant Biotechnology Journal* 14(1): 284–98.

Lin, HJ., HY. Cheng, CH. Chen, HC. Huang, and T-Y. Feng. "Plant amphipathic proteins delay the hypersensitive response caused by harpin(PSS) and *Pseudomonas syringae* pv. *syringae*." *Physiological and Molecular Plant Pathology* 51, no. 6 (1997): 367–76.

Lopez, Roberto G., and Erik S. Runkle. "Environmental physiology of growth and flowering of orchids." *HortScience* 40, no. 7 (2005): 1969–73.

Martin, Cathie, and Tom Gerats. "Control of pigment biosynthesis genes during petal development." *The Plant Cell* 5, no. 10 (1993): 1253.

Men, S., X. Ming, Y. Wang, R. Liu, C. Wei, and Y. Li. "Genetic transformation of two species of orchid by biolistic bombardment." *Plant Cell Reports* 21, no. 6 (2003): 592–98.

Men, Shuzhen, Xiaotian Ming, Rongwei Liu, Chunhong Wei, and Yi Li. "*Agrobacterium*-mediated genetic transformation of a *Dendrobium* orchid." *Plant Cell, Tissue and Organ Culture* 75, no. 1 (2003): 63–71.

Mishiba, Kei-ichiro, Dong Poh Chin, and Masahiro Mii. "*Agrobacterium*-mediated transformation of *Phalaenopsis* by targeting protocorms at an early stage after germination." *Plant Cell Reports* 24, no. 5 (2005): 297–303.

Motes, Martin R. *Vandas: their botany, history, and culture*. pp. 9–10. USA: Timber Press, 2004.

Nan, G-L., and A. R. Kuehnle. "Genetic transformation in *Dendrobium* (orchid)." In *Plant protoplasts and genetic engineering VI*, 145–155. Berlin, Heidelberg: Springer, 1995.

Nan, G-L., C-S. Tang, A. R. Kuehnle, and C. I. Kado. "*Dendrobium* orchids contain an inducer of *Agrobacterium* virulence genes." *Physiological and Molecular Plant Pathology* 51, no. 6 (1997): 391–99.

Nan, Guo-Ling, and Adelheid R. Kuehnle. "Factors affecting gene delivery by particle bombardment of *Dendrobium* orchids." *In Vitro Cellular & Developmental Biology-Plant* 31, no. 3 (1995): 131–36.

Nishihara, Masahiro, and Takashi Nakatsuka. "Genetic engineering of flavonoid pigments to modify flower color in floricultural plants." *Biotechnology Letters* 33, no. 3 (2011): 433–41.

Nogués, Isabel, Federico Brilli, and Francesco Loreto. "Dimethylallyl diphosphate and geranyl diphosphate pools of plant species characterized by different isoprenoid emissions." *Plant Physiology* 141, no. 2 (2006): 721–30.

Obara-Okeyo, P., and S. Kako. "Genetic diversity and identification of *Cymbidium* cultivars as measured by random amplified polymorphic DNA (RAPD) markers." *Euphytica* 99, no. 2 (1998): 95–101.

Okemura, Alan K. *Incidence and Exression of Cymbidium Mosaic Virus in Dendrobium Hybrids*. PhD diss., 1983.

Ono, Eiichiro, Masako Fukuchi-Mizutani, Noriko Nakamura, Yuko Fukui, Keiko Yonekura-Sakakibara, Masaatsu Yamaguchi, Toru

Nakayama, Takaharu Tanaka, Takaaki Kusumi, and Yoshikazu Tanaka. "Yellow flowers generated by expression of the aurone biosynthetic pathway." *Proceedings of the National Academy of Sciences* 103, no. 29 (2006): 11075–80.

Petroni, Katia, and Chiara Tonelli. "Recent advances on the regulation of anthocyanin synthesis in reproductive organs." *Plant Science* 181, no. 3 (2011): 219–29.

Qin, Xuyin, Yang Liu, Shanjing Mao, Tianbi Li, Hsinkan Wu, Chengcai Chu, and Youping Wang. "Genetic transformation of lipid transfer protein encoding gene in *Phalaenopsis amabilis* to enhance cold resistance." *Euphytica* 177, no. 1 (2011): 33–43.

Raffeiner, Barbara, Margrethe Serek, and Traud Winkelmann. "*Agrobacterium* tumefaciens-mediated transformation of *Oncidium* and *Odontoglossum* orchid species with the ethylene receptor mutant gene etr1-1." *Plant Cell, Tissue and Organ Culture (PCTOC)* 98, no. 2 (2009): 125–34.

Retheesh, S. T., and A. I. Bhat. "Simultaneous elimination of *Cucumber* mosaic virus and *Cymbidium* mosaic virus infecting *Vanilla planifolia* through meristem culture." *Crop Protection* 29, no. 10 (2010): 1214–17.

Sandmann, Gerhard, Susanne Römer, and Paul D. Fraser. "Understanding carotenoid metabolism as a necessity for genetic engineering of crop plants." *Metabolic Engineering* 8, no. 4 (2006): 291–302.

Sanford, John C., Theodore M. Klein, Edward D. Wolf, and Nelson Allen. "Delivery of substances into cells and tissues using a particle bombardment process." *Particulate Science and Technology* 5, no. 1 (1987): 27–37.

Sanjaya, X., and Ming-Tsair Chan. "Genetic transformation as a tool for improvement of orchids." In *Orchid Biotechnology*, 225–253, 2007.

Sawettalake, Nunchanoke, Sumontip Bunnag, Yanwen Wang, Lisha Shen, and Hao Yu. "DOAP1 promotes flowering in the orchid *Dendrobium* Chao Praya Smile." *Frontiers in Plant Science* 8 (2017): 400.

Searle, Iain, Yuehui He, Franziska Turck, Coral Vincent, Fabio Fornara, Sandra Kröber, Richard A. Amasino, and George Coupland. "The transcription factor FLC confers a flowering response to vernalization by repressing meristem competence and systemic signaling in *Arabidopsis*." *Genes & Development* 20, no. 7 (2006): 898–912.

Shi, N. N., Y. Xu, K. F. Yang, Y. Chen, H. Z. Wang, X. B. Xu, and Y. Hong. "Development of a sensitive diagnostic assay to detect *Cymbidium* mosaic virus and *Odontoglossum* ringspot virus in members of the Orchidaceae." *The Journal of Horticultural Science and Biotechnology* 86, no. 1 (2011): 69–73.

Shrestha, Bipna Rani, Dong Poh Chin, Ken Tokuhara, and Masahiro Mii. "Efficient production of transgenic plantls of *Vanda* through sonication-assisted *Agrobacterium*-mediated transformation of protocorm-like bodies." *Plant Biotechnology* 24, no. 4 (2007): 429–34.

Sjahril, Rinaldi, and Masahiro Mii. "High-efficiency *Agrobacterium*-mediated transformation of *Phalaenopsis* using meropenem, a novel antibiotic to eliminate Agrobacterium." *The Journal of Horticultural Science and Biotechnology* 81, no. 3 (2006): 458–64.

Sjahril, Rinaldi, Dong Poh Chin, Raham Sher Khan, Saburo Yamamura, Ikuo Nakamura, Yoshimiki Amemiya, and Masahiro Mii. "Transgenic *Phalaenopsis* plants with resistance to *Erwinia carotovora* produced by introducing wasabi defensin gene using *Agrobacterium* method." *Plant Biotechnology* 23, no. 2 (2006): 191–94.

Suwanaketchanatit, C., J. Piluek, S. Peyachoknagul, and P. S. Huehne. "High efficiency of stable genetic transformation in *Dendrobium* via microprojectile bombardment." *Biologia Plantarum* 51, no. 4 (2007): 720–27.

Suzuki, Atsushi, Midori Hayashida, Takeshi Ito, Hirokazu Kawano, Takeshi Nakano, Masayuki Miura, Kouichi Akahane, and Katsuya Shiraki. "Survivin initiates cell cycle entry by the competitive interaction with Cdk4/p16 INK4a and Cdk2/cyclin E complex activation." *Oncogene* 19, no. 29 (2000): 3225–34.

Tan, H. T. W. and Hew, C. S. *A guide of orchids of Singapore*. 2nd ed. Singapore Science Centre, Singapore, 1995.

Tanaka, Yoshikazu, and Akemi Ohmiya. "Seeing is believing: engineering anthocyanin and carotenoid biosynthetic pathways." *Current Opinion in Biotechnology* 19, no. 2 (2008): 190–97.

Tanaka, Yoshikazu, Filippa Brugliera, and Steve Chandler. "Recent progress of flower colour modification by biotechnology." *International Journal of Molecular Sciences* 10, no. 12 (2009): 5350–69.

Tanaka, Yoshikazu, Filippa Brugliera, Gianna Kalc, Mick Senior, Barry Dyson, Noriko Nakamura, Yukihisa Katsumoto, and Steve Chandler. "Flower color modification by engineering of the flavonoid biosynthetic pathway: practical perspectives." *Bioscience, Biotechnology, and Biochemistry* 74, no. 9 (2010): 1760–9.

Tang, Kexuan, Xiaofen Sun, Quanan Hu, Aizhong Wu, C-H. Lin, H-J. Lin, Richard M. Twyman, Paul Christou, and Tengyung Feng. "Transgenic rice plants expressing the ferredoxin-like protein (AP1) from sweet pepper show enhanced resistance to *Xanthomonas oryzae* pv. *oryzae*." *Plant Science* 160, no. 5 (2001): 1035–42.

Tee, C., M. Marziah, C. Tan, and M. Abdullah. "Evaluation of different promoters driving the GFP reporter gene and selected target tissues for particle bombardment of *Dendrobium* Sonia 17." *Plant Cell Reports* 21, no. 5 (2003): 452–58.

Tee, Chong Siang, and Mahmood Maziah. "Optimization of biolistic bombardment parameters for *Dendrobium* Sonia 17 calluses using GFP and GUS as the reporter system." *Plant Cell, Tissue and Organ Culture* 80, no. 1 (2005): 77–89.

Thiruvengadam, Muthu, Wei-Han Hsu, and Chang-Hsien Yang. "Phosphomannose-isomerase as a selectable marker to recover transgenic orchid plants (*Oncidium* Gower Ramsey)." *Plant Cell, Tissue and Organ Culture (PCTOC)* 104, no. 2 (2011): 239–46.

Thornberry, H. H., and M. R. Phillippe. "Orchid disease: *Cattleya* blossom brown necrotic streak." *Plant Disease Report* 48 (1964): 936–40.

Togami, Junichi, Hiroaki Okuhara, Noriko Nakamura, Kanako Ishiguro, Chika Hirose, Misa Ochiai, Yuko Fukui, Masa-atsu Yamaguchi, and Yoshikazu Tanaka. "Isolation of cDNAs encoding tetrahydroxychalcone 2′-glucosyltransferase activity from carnation, cyclamen, and catharanthus." *Plant Biotechnology* 28, no. 2 (2011): 231–38.

Tsuda, Shinzo, Yuko Fukui, Noriko Nakamura, Yukihisa Katsumoto, Keiko Yonekura-Sakakibara, Masako Fukuchi-Mizutani, Kazuyuki Ohira et al. "Flower color modification of *Petunia hybrida* commercial varieties by metabolic engineering." *Plant Biotechnology* 21, no. 5 (2004): 377–86.

Wang, H-M, K-Y To, H-M Lai, and S-T Jeng. "Modification of Flower Colour by Suppressing β-Ring Carotene Hydroxylase Genes in *Oncidium*." *Plant Biology* 18, no. 2 (2016): 220–29.

Wang, Shan-Li, Kotapati Kasi Viswanath, Chii-Gong Tong, Hye Ryun An, Seonghoe Jang, and Fure-Chyi Chen. "Floral induction and flower development of orchids." *Frontiers in Plant Science* 10 (2019): 1258.

Wen, Zhenzhen, Wenzhong Guo, Jinchi Li, Haisheng Lin, Chunmei He, Yunquan Liu, Qunyu Zhang, and Wei Liu. "Comparative transcriptomic analysis of vernalization-and cytokinin-induced floral transition in *Dendrobium nobile*." *Scientific Reports* 7, no. 1 (2017): 1–12.

Williams, Christine A., and Renee J. Grayer. "Anthocyanins and other flavonoids." *Natural Product Reports* 21, no. 4 (2004): 539–73.

Xiang, Lin, Xiaobai Li, Dehui Qin, Fangqi Guo, Chao Wu, Lixiang Miao, and Chongbo Sun. "Functional analysis of FLOWERING LOCUS T orthologs from spring orchid (*Cymbidium goeringii* Rchb. f.) that regulates the vegetative to reproductive transition." *Plant Physiology and Biochemistry* 58 (2012): 98–105.

Xiang, N., Y. Hong, and L. T. Lam-Chan. "Genetic analysis of tropical orchid hybrids (*Dendrobium*) with fluorescence amplified fragment-length polymorphism (AFLP)." *Journal of the American Society for Horticultural Science* 128, no. 5 (2003): 731–35.

Yang, J., H-J. Lee, D. H. Shin, S. K. Oh, J. H. Seon, K. Y. Paek, and K-H. Han. "Genetic transformation of *Cymbidium* orchid by particle bombardment." *Plant Cell Reports* 18, no. 12 (1999): 978–84.

Yee, Ng Chea, Janna Ong Abdullah, Maziah Mahmood, and Nazir Basiron. "Co-transfer of gfp, CHS and hptII genes into *Oncidium* Sharry Baby PLB using the biolistic gun." *African Journal of Biotechnology* 7, no. 15 (2008).

You, Su-Juan, Chia-Hui Liau, Hsiang-En Huang, Teng-Yung Feng, Venkatesh Prasad, Hsin-hao Hsiao, Jian-Cheng Lu, and Ming-Tsair Chan. "Sweet pepper ferredoxin-like protein (pflp) gene as a novel selection marker for orchid transformation." *Planta* 217, no. 1 (2003): 60–65.

Yu, H., S. H. Yang, and C. J. Goh. "*Agrobacterium*-mediated transformation of a *Dendrobium* orchid with the class 1 knox gene DOH1." *Plant Cell Reports* 20, no. 4 (2001): 301–05.

Yuen, C. K. K. H., H. Kamemoto, and M. Ishii. "Transmission of cymbidium mosaic virus through seed propagation in *Dendrobium*." *American Orchid Society Bulletin* 48, no. 12 (1979): 1245–47.

Zettler, F. W., N. J. Ko, G. C. Wisler, M. S. Elliott, and S. M. Wong. "Viruses of orchids and their control." *Plant Disease* 74, no. 9 (1990): 621–26.

Zhang, Jian-Xia, Kun-Lin Wu, Li-Ning Tian, Song-Jun Zeng, and Jun Duan. "Cloning and characterization of a novel CONSTANS-like gene from *Phalaenopsis hybrida*." *Acta Physiologiae Plantarum* 33, no. 2 (2011): 409–17.

Zhang, Lin, Dong Poh Chin, and Masahiro Mii. "Agrobacterium-mediated transformation of protocorm-like bodies in *Cattleya*." *Plant Cell, Tissue and Organ Culture (PCTOC)* 103, no. 1 (2010): 41–47.

Zheng, Q., Y. P. Zheng, G. D. Wang, W. M. Guo, W. F. Fan, and C. Wang. "Sonication-assisted *Agrobacterium*-mediated transformation of the ACC gene to interfere the production of ethylene in spring *Dendrobium* cv.'Sanya.'" *Russian Journal of Plant Physiology* 59, no. 2 (2012): 266–74.

Zheng, You-Xiu, Bing-Nan Shen, Ching-Chung Chen, and Fuh-Jyh Jan. "*Odontoglossum* ringspot virus causing flower crinkle in *Phalaenopsis* hybrids." *European Journal of Plant Pathology* 128, no. 1 (2010): 1–5.

Zheng, You-Xiu, Ching-Chung Chen, Yuh-Kun Chen, and Fuh-Jyh Jan. "Identification and characterization of a potyvirus causing chlorotic spots on *Phalaenopsis* orchids." *European Journal of Plant Pathology* 121, no. 1 (2008): 87–95.

About the Editor

Vladan Djordjević, PhD
Research Associate,
Faculty of Biology, Institute of Botany and Botanical Garden,
University of Belgrade, Serbia
vdjordjevic@bio.bg.ac.rs

Vladan Djordjević is a Research Associate in the Department of Ecology and Phytogeography, Institute of Botany and Botanical Garden, Faculty of Biology, University of Belgrade. He graduated from the Faculty of Biology, University of Belgrade, and later earned a PhD degree from the same faculty. Since 2015 he has been working at the Institute of Botany and Botanical Garden, Faculty of Biology, University of Belgrade, Serbia. His research focus is on ecology, distribution, and conservation of terrestrial orchids.

INDEX

A

B

C

D

E

F

G

H

I

J

K

L

O

P

R

S

T

U

V

W

Y

Z